Female Infidelity and Paternal Uncertainty

Although it is commonly believed that males are more promiscuous than females, new research has revealed the frequency of female infidelity and the consequences of such behavior. Because males cannot be certain of the paternity of their putative offspring, males have evolved a number of anti-cuckoldry strategies to deal with the potential possibility of raising an offspring they unknowingly did not sire. *Female Infidelity and Paternal Uncertainty: Evolutionary Perspectives on Male Anti-cuckoldry Tactics* is the first book to address these perspectives and look at how males deal with the consequences of female infidelity. Each chapter deals with a specific evolved strategy developed to aid males in either limiting opportunities for their mate to be unfaithful or to 'correct' the byproducts of infidelity should it occur. With sections including mate guarding, intravaginal tactics, and paternity assessment, this book will appeal to researchers and graduate students in behavioral biology, evolutionary psychology, human sexuality, anthropology, sociology, reproductive health, and medicine.

STEVEN M. PLATEK is Assistant Professor of Psychology and Biomedical Science at Drexel University in Philadelphia and director of the Evolutionary Cognitive Neuroscience Laboratory, which applies cognitive neuroscience methods (e.g., fMRI, optical imaging) to the study of parental behavior, sex differences, deception, and self-awareness.

TODD K. SHACKELFORD is Associate Professor of Psychology, Graduate Program Co-Director, and Chair of the Evolutionary Psychology Area of Florida Atlantic University. Dr. Shackelford directs the Evolutionary Psychology Laboratory, which uses a modern evolutionary psychological perspective to investigate social and interpersonal phenomena.

Female Infidelity and Paternal Uncertainty

Evolutionary Perspectives on Male Anti-cuckoldry Tactics

Edited by

STEVEN M. PLATEK

Drexel University
Philadelphia, USA

TODD K. SHACKELFORD

Florida Atlantic University
2912 College Avenue
Davie, USA

CAMBRIDGE UNIVERSITY PRESS
Cambridge, New York, Melbourne, Madrid, Cape Town, Singapore, São Paulo

Cambridge University Press
The Edinburgh Building, Cambridge CB2 2RU, UK

Published in the United States of America by Cambridge University Press, New York

www.cambridge.org
Information on this title: www.cambridge.org/9780521845380

First published 2006

Printed in the United Kingdom at the University Press, Cambridge

A catalog record for this publication is available from the British Library

ISBN-13 978-0-521-84538-0 hardback
ISBN-10 0-521-84538-6 hardback
ISBN-13 978-0-521-60734-6 paperback
ISBN-10 0-521-60734-5 paperback

Contents

Contributors

Rebecca L. Burch
Department of Psychology, State University of New York at Oswego, Oswego, NY 13126, USA

Jennifer A. Davis
Department of Psychology, State University of New York at Albany, SUNY, 1400 Washington Avenue, Albany, NY 12222, USA

Gordon G. Gallup, Jr.
Department of Psychology, State University of New York at Albany, SUNY, 1400 Washington Avenue, Albany, NY 12222, USA

Steven W. Gangestad
Department of Psychology, University of New Mexico, Albuquerque, NM 87111, USA

David C. Geary
Department of Psychological Sciences, University of Missouri – Columbia, Columbia, MO 65211-2500, USA

Aaron T. Goetz
Department of Psychology, Florida Atlantic University, 2912 College Avenue, Davie, FL 33314, USA

Daniel Hipp
State University of New York at Oswego, 7060 Route 104, Oswego, NY 13126-3599, USA

Steven M. Platek
Drexel University, Department of Psychology, Philadelphia, PA 19102, USA

Todd K. Shackelford
Florida Atlantic University, Department of Psychology, 2912 College Avenue, Davie, FL 33314, USA

Jaime W. Thomson
Department of Psychology, Drexel University, 3141 Chestnut Street, Philadelphia, PA 19104, USA

Acknowledgments

Steven M. Platek thanks, for scholarly support, encouragement, and discussion about this research, Gordon Bear, Rebecca Burch, Martin Daly, Jennifer Davis, Scott Faro, Ludivine Fonteyn, Gordon Gallup, Jr., Steven Gaulin, Aaron Goetz, Ruben Gur, Robert Haskell, Farzin Irani, Julian Paul Keenan, Daniel Langleben, Sarah Levin, James Loughead, Rick Michalski, Feroze Mohamed, Thomas Myers, Ivan Panyavin, Shilpa Patel, Katie Rodak, Michele Sackowicz, David Smith, Jaime Thomson, and Margo Wilson. Special thanks to my co-editor, Todd Shackelford, for enthusiasm and dedication to this volume without which this volume may not have been. Also, special thanks to Patricia and Joesph Platek, for their continued support of my academic and personal endeavours.

Todd K. Shackelford thanks, for their scholarly support and encouragement, John Alcock, Robin Baker, Mark Bellis, Iris Berent, Jesse Bering, Tim Birkhead, Dave Bjorklund, April Bleske-Rechek, Rebecca Burch, David Buss, Martin Daly, Harald Euler, Gordon Gallup, Steve Gangestad, Aaron Goetz, Martie Haselton, Steve Hecht, Erika Hoff, Sabine Hoier, Lee Kirkpatrick, Craig LaMunyon, Randy Larsen, Brett Laursen, Rick Michalski, Dave Perry, Steven Platek, Nick Pound, Monica Rosselli, Dave Schmitt, Bob Smith, Randy Thornhill, Robin Vallacher, Charles White, and Margo Wilson. Special thanks to Steven Platek, for his hard work and persistence in bringing this volume to fruition. Finally, my deepest thanks to Viviana Weekes-Shackelford, for her unwavering support and encouragement, professional and personal.

The editors thank Martin Griffiths at Cambridge University Press for his direction, support, and patience.

PART I   INTRODUCTION AND OVERVIEW

1

Introduction to theory and research on anti-cuckoldry tactics: overview of current volume

STEVEN M. PLATEK
Drexel University
AND
TODD K. SHACKELFORD
Florida Atlantic University

Female infidelity

In most cultures, marriage vows entail the promise of fidelity and life-long commitment. In principle, marriage vows are a contract – a reproductive contract – between two individuals to maintain both emotional *and* sexual fidelity to one another 'til death do them part.' Monogamy. There are few species that maintain monogamous relationships between the sexes. It is commonly believed that males are more promiscuous, but new research is shedding light on the prominence of female infidelity as well as the consequences of such behavior.

Female infidelity is common in the animal kingdom as well as among humans. According to an analysis of 280 000 paternity tests conducted in 1999 by the American Association of Blood Banks, approximately 30% of children are fathered by extra-pair copulations; that is, 30% of children in this sample were fathered by someone other than the woman's long-term romantic partner. Several case studies exemplify this phenomenon and the associated psychological and social consequences. The *New York Times* reported a case of a Texas man who was faced with the unnerving news that not one, but several, of his children were the product of extra-pair paternity. The bittersweet news came when the man was being tested as a carrier for a debilitating genetic disorder that his youngest daughter had suffered with since birth. When the genetic test came back negative he should have been elated, but knowing that both parents must be carriers for any child to be inflicted with the disorder raised obvious concerns. Either the doctor misdiagnosed his child or his child was the product of infidelity on the part of his wife. DNA paternity testing confirmed that he did not father his youngest daughter. This motivated him to obtain paternity tests for his three other children, only one of which had been sired by him.

A team of scientists in Italy (Barbaro *et al.*, 2004) was asked to conduct DNA paternity tests to reveal relatedness among individuals involved in a murder case. The case involved three victims (two males, one female) that were brutally murdered. The suspected killer was later found hanged, a presumed suicide. The police conducted a series of DNA paternity tests to determine the paternity of the murdered female's child. It was presumed that the child had been fathered by the suspect, the man who had hanged himself, because he had been in a long-term relationship with the slain female. However, DNA paternity testing revealed that the child was actually fathered by one of the other murdered young men. A judge in another case ordered a team of scientists to solve a disputed paternity case among a set of twin girls when one of the girls raised doubts about her father's paternity, because she felt the man favored her twin sister. DNA testing confirmed that the man had only sired one of the twin girls and was genetically excluded from having sired the twin who raised doubts about paternity. It is not uncommon in the animal literature to find organisms that will have offspring (of a single litter) fathered by several males. For example, in most passerine birds, any given clutch can contain eggs fathered by two or more males. However, this phenomenon has rarely been described or discussed in humans.

The problem of paternal uncertainty: how males deal with female infidelity

Because of concealed ovulation, internal fertilization, and female infidelity, human parental certainty is asymmetrical: unlike females, who are always 100% certain of *maternity*, males can never be certain of *paternity*. Current estimates of extra-pair paternity (paternity by someone other than the putative and domestic father, or cuckoldry) are between 1 and 30%, with the best estimate at about 10% (Baker & Bellis, 1995; Cerda-Flores *et al.*, 1999; Neale, Neale, & Sullivan, 2002; Sasse *et al.*, 1994; Sykes & Irven, 2000). In other words, approximately 1 in 10 children are the product of female infidelity. This asymmetry in parental certainty has contributed to an asymmetry in human parental investment (Bjorklund & Shackelford, 1999; Dawkins, 1976; Geary, 2000; Trivers, 1974). As a consequence of having to carry a child to term, females, by default, invest more in and provision more for children than do males. Additionally, if a female nurses her offspring she could be bound to several additional years of parental investment.

Males, however, are not obligated to invest resources in offspring, and tend to provide care proportional to their confidence or certainty of paternity (Burch & Gallup, 2000; Daly & Wilson, 1996, 1998). The risk of cuckoldry appears to have driven the evolution of male anti-cuckoldry tactics; tactics designed to limit and control female infidelity in an attempt to reduce the risk of cuckoldry (Buss, 1988, 1994, 1999; Buss & Shackelford, 1997; Platek, 2002; Platek et al., 2002, 2003; Shackelford et al., 2002).

We can observe a similar pattern among many other mammals. For example, paternal care is usually only manifest in those mammals with relatively high paternal certainty, whereas for most mammals (95–97%) males provide little or no direct investment in their offspring. For those few species that do engage in paternal provisioning, it appears that the males have evolved several anti-cuckoldry tactics that increase the certainty that they are the source of paternity (e.g. Lacy & Sherman, 1983). In an attempt to limit provisioning for offspring that are the consequence of female extra-pair copulations, males of some species sometimes perform what may appear to be extreme behaviors. For example, when a male langur overthrows another male and gains dominance within a troop, he will systematically kill infants that were fathered by the previous alpha male. By resorting to infanticide when the paternity of an offspring is ostensibly foreign (e.g. Hrdy, 1974), his behavior serves two functions: (1) it eliminates the possibility that he will invest valuable resources in unrelated offspring and (2) it induces sexual receptivity in those females whose offspring he killed. This allows the new dominant male to use the females for his own reproductive best interests. Additionally, male baboons have been found to invest resources in offspring proportional to the degree to which he monopolized the females prior to insemination (Buchan et al., 2003).

There is growing evidence that human males are similarly affected by these evolutionary pressures to invest in offspring as a function of paternal certainty. As a way of elucidating the importance of paternity for males, Daly and Wilson (1982) and Regalski and Gaulin (1993) (see also Brédart & French, 1999; Christenfeld & Hill, 1995; Nesse, Silverman, & Bortz, 1990) observed family interactions following the birth of a child. They measured the number of times people remarked who the infant looked like. Both studies found that people were more likely to comment that the children resembled the putative fathers than they were to comment that the children resembled the mothers. Both studies also documented that the mother and her family were more likely to attribute resemblance to the putative father than to the mother, whereas the putative father and his family showed no such bias. The authors interpreted these behaviors as attempts on the part of the female and her family to reassure the putative father and his family about paternity. Daly and Wilson (1982) report that one male was so adamant about the importance of paternity that he stated, "... if the child does not look like me I'll abandon them both!"

It is also well known that men preferentially invest resources in children to whom they are likely to be related genetically. For example, it is not uncommon for stepchildren to be treated significantly worse than genetic children (e.g. Anderson *et al.*, 1999). Burch and Gallup (2000) have shown that males spend less time with, invest fewer resources in, and are more likely to abuse ostensibly unrelated children than children they assume to be their genetic offspring. They also found that the less a male thinks a child (unrelated or genetic) looks like him, the worse he treats the child and the worse he views the relationship with that child. Daly and Wilson (1988; and see Daly, Wilson, & Weghorst, 1982) estimate the incidence of filicide (child-killing) among stepchildren to be as much as 100 times that of genetic children. In Daly and Wilson's (1988) landmark book, *Homicide*, they interpret spousal homicide as a byproduct of cuckoldry fear and sexual jealousy among men. These data suggest a strong link between paternity uncertainty and the treatment of children.

As a result of paternal uncertainty, human males appear to have evolved an arsenal of anti-cuckoldry tactics to limit and perhaps control the incidence of female infidelity and thereby increase the likelihood that the children they provision are genetically their own. Emerging data suggest that males have evolved at least three types of tactic that help to reduce the likelihood of being cuckolded. Each of these tactics is the focus of a section of this volume, with chapters contributed by leading experts in the field of evolutionary science.

This volume takes the form of an integration of new data with newly emerging theory about human paternal uncertainty and the evolution of male anti-cuckoldry tactics in an attempt to consolidate a base of literature into a new model for the evolution of male anti-cuckoldry tactics. The conception of this volume was the result of a symposium, chaired by the editors, at the 15th annual meeting of the Human Behavior and Evolution Society at the University of Nebraska-Lincoln in June 2003. Dr. David Buss, Dr. Steven Gangestad, and Dr. Randy Thornhill, internationally renowned experts in evolutionary psychology, served as discussants to this symposium and Dr. Gangestad has contributed a chapter to the volume. The book consists of both papers presented as part of the symposium and other independent contributions that were not presented as part of the symposium, but represent significant advances in the relevant fields.

The first anti-cuckoldry tactic (addressed in Part II) involves attempts by a man to limit his mate's opportunities for extra-pair copulations that could result in pregnancy. Males have evolved psychological adaptations that produce mate guarding and jealousy (Buss, 1988; Buss & Shackelford, 1997; see Buss, 2000, for a review) to reduce or to prevent a mate from being inseminated by another male. Recent evidence suggests that males maximize the utility of their mate-guarding strategies by implementing them at ovulation, a key reproductive time in a female's menstrual cycle (Gangestad, Thornhill, & Garver, 2002). Further, jealousy appears to fluctuate with a man's mate value and, hence, risk of cuckoldry. Brown and Moore (2003), for example, found that males who were less symmetrical were significantly more jealous. These and other data suggest that jealousy has evolved as a means by which males can attempt to deter extra-pair copulations (Buss, 2000; Daly & Wilson, 1982; see also Buss et al., 2000; Buss & Shackelford, 1997). Chapters 3–5 address events that might lead up to female infidelities and the mate-guarding tactics employed to deter these behaviors. In Chapter 3, Steven Gangestad provides a discussion of the evidence for adaptations for female extra-pair mating in humans, how female extra-pair mating can lead to extra-pair paternity, and the likelihood that these behaviors are being driven by a female's search for "good genes" – genes that provide an advantage to offspring in the form of pathogen resistance and developmental stability.

In Chapter 4, Todd Shackelford and Aaron Goetz discuss commitment, devotion, and other mate-retention tactics as predictors of violence against women. Male sexual jealousy is a primary cause of violence in romantic relationships (Buss, 2000; Daly & Wilson, 1988). Mate-retention tactics are behavioral manifestations of jealousy designed to thwart a partner's infidelities or relationship defection (Buss, 1988; Buss & Shackelford, 1997; Flinn, 1988). Although some mate-retention tactics appear to be innocuous romantic gestures, some may be harbingers of violence. Shackelford and colleagues investigated the relationships between male mate-retention tactics and female-directed violence in romantic relationships. In an initial study, men reported their use of mate-retention tactics and violence in romantic relationships. Because research has shown that men underreport the violence they inflict on their partners, whereas women report this violence with relative accuracy, a second study assessed women's reports of their partners' behaviors. As predicted, and across both studies, men's use of particular mate-retention tactics was related positively to female-directed violence. A third study secured husbands' reports of their own mate retention and wives' reports of their husbands' use of violence. Again, men's use of particular mate-retention tactics and female-directed violence were related positively. Shackelford and colleagues conclude with a discussion of mate-retention tactics as unique predictors of violence in romantic relationships and suggest directions for future work. Concluding Part II

(Chapter 5), Goetz and Shackelford address the issue of forced in-pair copulation as an anti-cuckoldry tactic and provide interesting new data to support their theoretical model.

In Part III (Chapters 6–10), intra-vaginal anti-cuckoldry strategies such as sperm competition and semen displacement are discussed. In species with internal fertilization, sperm competition occurs when the sperm of two or more males simultaneously occupy the reproductive tract of a female and compete to fertilize an egg (Baker & Bellis, 1995; Parker, 1970, 1984). Sperm-competition theory provides the theoretical framework for a body of work investigating anatomical, behavioral, and psychological adaptations in males and females designed to solve problems posed by sperm competition (Parker, 1970; Smith, 1984). Although much current research on the evolutionary causes and consequences of sperm competition focus primarily on birds (Birkhead & Møller, 1992) and insects (Cooke & Gage, 1995; Gage, 1991; Thornhill & Alcock, 1983), studies of human sperm competition have become a recent focus by evolutionary biologists and evolutionary psychologists (Baker & Bellis, 1988, 1989a, 1989b, 1993a, 1993b, 1995; Bellis & Baker, 1990; Gangestad & Thornhill, 1997, 1998; Gangestad, Thornhill, & Garver, 2002; Pound, 2002; Shackelford & LeBlanc, 2001; Shackelford et al., 2000, 2002; Singh et al., 1998; Thornhill, Gangestad, & Comer, 1995).

Baker and Bellis (1993a, 1995), for example, documented that male humans, like male birds, male insects, and other male non-human primates, appear to be physiologically designed to solve the adaptive problems of sperm competition. Baker and Bellis studied couples in committed, sexual relationships and reported that, consistent with sperm-competition theory, when copulating with their partner men inseminate more sperm when the risk of sperm competition is high. Specifically, controlling for the time since last ejaculation, they documented a large positive correlation between the number of sperm ejaculated and the proportion of time a couple has spent apart since their last copulation.

In a study modeled after research by Baker and Bellis (1993a) on male-ejaculate adjustment as a function of the risk of female infidelity, Shackelford et al. (2002) found psychological evidence suggesting a long evolutionary history of human sperm competition. The ejaculate adjustment documented by Baker and Bellis (1993a, 1995) would not be functional if men were not motivated to copulate with their partners sooner rather than later following the possibility of rival insemination. Shackelford et al. (2002) documented a positive relationship between the proportion of time a couple has spent apart since their last copulation and, for example, men's ratings of their partner's sexual attractiveness and men's ratings of their interest in copulating with their partner.

Opening up Part III (Chapter 6) Aaron Goetz and Todd Shackelford introduce intra-vaginal anti-cuckoldry tactics with a thorough review of the sperm-competition literature. They synthesize existing and new data that suggest that human (and animal) male physiology adjusts the delivery of sperm quantity and semen chemistry as a function of cuckoldry risk and that there may be specific psychological adaptations associated with sperm competition.

In Chapter 7, Gordon Gallup and Rebecca Burch introduce the semen-displacement hypothesis, which suggests that the morphology of the human penis may act to plunge another male's ejaculate from the reproductive tract of his partner. Gallup and Burch (2004) documented that men's and women's perceptions of a male's sexual behavior (e.g. depth, vigor, and speed of penile thrusting) change as a function of cuckoldry risk. Additionally, by utilizing prosthetic male and female genitalia, they provide evidence that the morphology of the human penis, specifically the glans penis, the frenulum, and the coronal ridge, may be designed for successful semen displacement. This chapter capitulates this hypothesis and raises some interesting predictions based on their findings. Burch and Gallup also contribute a chapter (Chapter 8) that discusses a series of working hypotheses that involve the effects of semen chemistry and it's psychobiological effects as an intra-vaginal means by which males may be able to reduce the likelihood of cuckoldry or alter the hormonal status of a reproductive partner. This chapter provides great detail about the chemical composition of semen and the impact this cocktail may have on human mating patterns.

In Chapter 9, Aaron Goetz and Todd Shackelford describe how semen displacement, sperm competition, and mate-retention tactics represent a cluster of psychological adaptations that have all evolved to deal with female extra-pair mating. Goetz and colleagues have discovered that the use of mate-retention tactics (e.g. Buss, 1988; including vigilance, emotional manipulation, verbal and physical possession signals, and violence against rivals) correlates with the likelihood of sperm competition and, consequently, copulatory behaviors designed to displace rival male semen. Jennifer Davis and Gordon Gallup conclude Part III (Chapter 10) with a chapter that outlines a new hypothesis suggesting that the reproductive endocrinological disorder known as preeclampsia, an immune disorder that results from genetic incompatibilities between the parents, may have been selected for as a response in females to unfamiliar semen.

Part IV consists of two chapters and introduces the last in this line of cuckoldry defenses: assessing paternity post-parturition. It has been hypothesized that one way a male ancestor assessed paternity was by assessing the degree to which a child resembled him (Daly & Wilson, 1982, 1998; Regalski & Gaulin, 1993). In Chapter 11 Rebecca Burch and her colleagues summarize the "social mirror" effect; the degree to which social perceptions of paternal resemblance impact parental/paternal behaviors. Finally, in Chapter 12 Steven Platek and Jaime Thomson summarize the evidence that facial resemblance may act as a cue to kinship and, in particular, a cue to paternity for males. They summarize recent findings demonstrating unique neurobiological correlates for processing facial resemblance and the expression of sex differences. They also present evidence that males use self-child resemblance in their decisions to provision for offspring. Using facial morphing, Platek *et al.* (2002, 2003) have shown that males, but not females, respond favorably to facial resemblance in children when queried about hypothetical investment behaviors (e.g. which child would you spend the most time with?). Using functional magnetic resonance imaging (fMRI), Platek and his colleagues have shown that the brains of males react differentially to children's faces that resemble them. Male brain activity displayed when looking at children's faces that do not resemble their own does not differ from the brain activity displayed by females viewing faces of children, and females do not show differential activation as a function of self-child facial resemblance. These data suggest that males have evolved specific neurocognitive mechanisms that influence their decisions to provision for children.

This volume represents the inter-disciplinary and integrated approach to the study of paternal uncertainty and should shed new light on the topic both from a research and an applied perspective. This new three-stage theory – (1) mate-guarding strategies, (2) intra-vaginal strategies, and (3) post-partum strategies – generates many new testable hypotheses that we hope will further our understanding of female infidelity and anti-cuckoldry tactics.

References

Anderson, K., Kaplan, H., Lam, D., and Lancaster, J. (1999). Paternal care of genetic fathers and stepfathers II: reports by Xhosa high school students. *Evolution and Human Behavior*, **20**, 433–51.

Baker, R. R. and Bellis, M. A. (1988). "Kamikaze" sperm in mammals? *Animal Behaviour*, **36**, 936–9.

Baker, R. R. and Bellis, M. A. (1989a). Number of sperm in human ejaculates varies in accordance with sperm competition theory. *Animal Behaviour*, **37**, 867–9.

Baker, R. R. and Bellis, M. A. (1989b). Elaboration of the kamikaze sperm hypothesis: a reply to Harcourt. *Animal Behaviour*, **37**, 865–7.

Baker, R. R. and Bellis, M. A. (1993a). Human sperm competition: ejaculate adjustment by males and the function of masturbation. *Animal Behaviour*, **46**, 861–85.

Baker, R. R. and Bellis, M. A. (1993b). Human sperm competition: ejaculate manipulation by females and a function for the female orgasm. *Animal Behaviour*, **46**, 887–909.

Baker, R. R. and Bellis, M. A. (1995). *Human Sperm Competition: Copulation, Masturbation, and Infidelity*. London: Chapman and Hall.

Barbaro, A., Cormaci, P., Barbaro, A., and Louahlia, S. (2004). DNA analysis in a case of serial murders. *International Congress Series*, **1261**, 465–7.

Bellis, M. A. and Baker, R. R. (1990). Do females promote sperm competition: data for humans. *Animal Behavior*, **40**, 197–9.

Birkhead, T. R. and Møller, A. P. (1992). *Sperm Competition in Birds*. London: Academic Press.

Bjorklund, D. F. and Shackelford, T. K. (1999). Differences in parental investment contribute to important differences between men and women. *Current Directions in Psychological Science*, **8**(3), 86–9.

Brédart, S. and French, R. (1999). Do babies resemble their fathers more than their mothers? A failure to replicate Christenfeld and Hill. *Evolution and Human Behavior*, **20**, 129–35.

Brown, W. and Moore, C. (2003) Fluctuating asymmetry and romantic jealousy. *Evolution and Human Behavior*, **24**, 113–17.

Buchan, J. C., Alberts, S. C., Silk, J. B., and Altmann, J. (2003). True paternal care in a multi-male primate society. *Nature*, **425**, 179–81.

Burch, R. L. and Gallup, Jr., G. G. (2000). Perceptions of paternal resemblance predict family violence. *Evolution and Human Behavior*, **21**(6), 429–35.

Buss, D. M. (1988). From vigilance to violence: tactics of mate retention in American undergraduates. *Ethology and Sociobiology*, **9**, 291–317.

Buss, D. M. (1994). *The Evolution of Desire*. New York: Basic Books.

Buss, D. M. (1999). *Evolutionary Psychology*. Needham Heights, MA: Allyn & Bacon.

Buss, D. M. (2000). *The Dangerous Passion*. New York: Free Press.

Buss, D. M. and Shackelford, T. K. (1997). From vigilance to violence: mate retention tactics in married couples. *Journal of Personality and Social Psychology*, **72**, 346–61.

Buss, D. M., Shackelford, T. K., Choe, J., Buunk, B. P., and Dijkstra, P. (2000). Distress about mating rivals. *Personal Relationships*, **7**, 235–43.

Cerda-Flores, R. M., Barton, S. A., Marty-Gonzales, L. F., Rivas, F., and Chakraborty, R. (1999). Estimation of nonpaternity in the Mexican population of Nuevo Leon: a validation study with blood group markers. *American Journal of Physical Anthropology*, **109**, 281–93.

Christenfeld, N. and Hill, E. (1995). Whose baby are you? *Nature*, **378**, 669.

Cooke, P. A. and Gage, M. J. G. (1995) Effects of different risks of sperm competition upon eupyrene and apyrene sperm numbers in the moth *Plodia interpunctella*. *Behavioral Ecology and Sociobiology*, **36**, 261–8.

Daly, M. and Wilson, M. (1982). Whom are newborn babies said to resemble? *Ethology and Sociobiology*, **3**, 69–78.

Daly, M. and Wilson, M. (1988). *Homicide*. New York: Adeline de Gruyter.

Daly, M. and Wilson, M. I. (1996). Violence against stepchildren. *Current Directions in Psychological Science*, **5**, 77–81.

Daly, M. and Wilson, M. (1998). *The Truth about Cinderella: a Darwinian View of Parental Love*. New Haven, CT: Yale University Press.

Daly, M., Wilson, M., and Weghorst, S. J. (1982). Male sexual jealousy. *Ethology and Sociobiology*, **3**(1), 11–27.

Dawkins, R. (1976). *The Selfish Gene*. New York: Oxford University Press.

Flinn, M. (1988). Mate guarding in a Caribbean village. *Ethology and Sociobiology*, **9**, 1–28.

Gage, M. J. G. (1991) Sperm competition risk directly affects ejaculate size in the Mediterranean fruit fly. *Animal Behaviour*, **42**, 1036–7.

Gallup G. G. and Burch, R. L. (2004). Semen displacement as a sperm competition strategy in humans. *Evolutionary Psychology*, **2**, 12–23.

Gangestad, S. W. and Thornhill, R. (1997). The evolutionary psychology of extra-pair sex: the role of fluctuating asymmetry. *Evolution and Human Behavior*, **18**, 69–88.

Gangestad, S. W. and Thornhill, R. (1998). Menstrual cycle variation in women's preferences for the scent of symmetrical men. *Proceedings of the Royal Society of London B*, **265**, 927–33.

Gangestad, S. W., Thornhill, R., and Garver, C. E. (2002). Changes in women's sexual interests and their partners' mate retention tactics across the menstrual cycle: evidence for shifting conflicts of interest. *Proceedings of the Royal Society of London B*, **269**, 975–82.

Geary, D. C. (2000). Evolution and proximate expression of human paternal investment. *Psychological Bulletin*, **126**, 55–77.

Hrdy, S. (1974). Male-male competition and infanticide among the langurs (*Presbytis entellus*) of Abu, *Folia Rajasthan. Primatologica*, **22**, 19–58.

Lacy, R. C. and Sherman, P. W. (1983). Kin recognition by phenotype matching. *American Naturalist*, **121**, 489–512.

Neale, M. C., Neale, B. M., and Sullivan, P. F. (2002). Nonpaternity in linkage studies of extremely discordant sib pairs. *American Journal of Human Genetics*, **70**, 526–9.

Nesse, R., Silverman, A., and Bortz, A. (1990) Sex differences in ability to recognize family resemblance. *Ethology and Sociobiology*, **11**, 11–21.

Parker, G. A. (1970). Sperm competition and its evolutionary consequences in the insects. *Biological Review*, **45**, 525–67.

Parker, G. A. (1984). Sperm competition and the evolution of animal mating strategies. In R. L. Smith, ed., *Sperm Competition and the Evolution of Animal Mating Systems*. NY: Academic Press, pp. 1–60.

Platek, S. M. (2002). Unconscious reactions to children's faces: the effect of resemblance. *Evolution and Cognition*, **8**, 207–14.

Platek, S. M., Burch, R. L., Panyavin, I. S., Wasserman, B. H., and Gallup, Jr., G. G. (2002). Reactions to children's faces: resemblance affects males more than females. *Evolution and Human Behavior*, **23**(3), 159–66.

Platek, S. M., Critton, S. R., Burch, R. L., Frederick, D. A., Myers, T. S. and Gallup, Jr., G. G. (2003) How much resemblance is enough? Determination of a just noticeable difference at which male reactions towards children's faces change from indifferent to positive. *Evolution and Human Behavior*, **23**, 81–7.

Pound, N. (2002). Male interest in visual cues of sperm competition risk. *Evolution and Human Behavior*, **23**, 443–66.

Regalski, J. and Gaulin, S. (1993). Whom are Mexican infants said to resemble? Monitoring and fostering paternal confidence in the Yucatan. *Ethology and Sociobiology*, **14**, 97–113.

Sasse, G., Muller, H., Chakraborty, R., and Ott, J. (1994). Estimating the frequency of nonpaternity in Switzerland. *Human Heredity*, **44**, 337–43.

Shackelford, T. K. and LeBlanc, G. J. (2001). Sperm competition in insects, birds, and humans: insights from a comparative evolutionary perspective. *Evolution and Cognition*, **7**, 194–202.

Shackelford, T. K., Weekes, V. A., LeBlanc, G. J., Bleske, A. L., Euler, H. A., and Hoier, S. (2000). Female coital orgasm and male attractiveness. *Human Nature*, **11**, 299–306.

Shackelford, T. K., LeBlanc, G. J., Weekes-Shackelford, V. A., *et al.* (2002). Psychological adaptation to human sperm competition. *Evolution and Human Behavior*, **23**, 123–38.

Singh, D., Meyer, W., Zambarano R., and Hurlbert, D. (1998). Frequency and timing coital orgasm in women desirous of becoming pregnant. *Archives of Sexual Behavior*, **27**(1), 15–29.

Smith, R. L. (1984). Human sperm competition. In R. L. Smith, ed., *Sperm Competition and the Evolution of Animal Mating Systems*. New York: Academic Press, pp. 601–60.

Sykes, B. and Irven, C. (2000). Surnames and the Y chromosome. *American Journal of Human Genetics*, **66**, 1417–19.

Thornhill, R. and Alcock, J. (1983). *The Evolution of Insect Mating Systems*. Cambridge, MA: Harvard University Press.

Thornhill, R., Gangestad, S. W., and Comer, R. (1995). Human female orgasm and mate fluctuating asymmetry. *Animal Behaviour*, **50**, 1601–15.

Trivers, R. L. (1974). Parent-offspring conflict. *American Zoologist*, **14**, 249–64.

2

Coevolution of paternal investment and cuckoldry in humans

DAVID C. GEARY
University of Missouri – Columbia

Introduction

The social dynamics of reproduction emerge as individuals compete for control of the reproductive potential and reproductive investment of members of the opposite sex. Reproductive potential is the individual's ability to invest in the growth, development, and later social and reproductive competencies of offspring and the potential genetic benefits that might be passed to offspring (Alexander, 1987; Geary, 2002). Reproductive investment is the expenditure of the behavioral components (e.g. provisioning) of this potential to enhance to the survival and later reproductive prospects of one offspring at the expense of other offspring or the parent (Trivers, 1972). As identified by Darwin (1871), the dynamics of reproduction – termed sexual selection – take the form of *intrasexual competition* for access to mates or for control of the resources that potential mates need to reproduce, and *intersexual choice* of mating partners (see also Andersson, 1994). For most species of mammal, intrasexual competition takes the form of physical contests, whereby males attempt to exclude other males from access to females or access to the resources that females need to reproduce. Intersexual choice often involves female choice of mates, based on the results of male–male competition or on indicators of male genetic fitness or the males' ability and willingness to invest reproductive potential in her offspring.

To the extent that reproductive interests of males and females differ, sexual selection will also involve *intersexual conflict*, which traditionally involves conflict over the allocation of parental investment (Haig, 1993; Trivers, 1972). Intersexual conflict will be most extreme for species in which reproduction involves prolonged male and female interdependence and when one or either sex can benefit by diverting reproductive investment from the core relationship to alternative relationships. Diversion of reproductive investment is of course related to intersexual choice and intrasexual competition. As an example, cuckoldry – whereby the male social partner is deceived into raising the offspring of another male – involves female choice but is complicated by the simultaneous need to maintain the relationship with the social partner while engaging in a sexual and potentially reproductive relationship with an extra-pair male. Among other things, this dynamic creates conditions that will favor the evolution of female strategies that function to deceive the primary male social partner into investing his reproductive potential in offspring that may be those of another male (i.e. cuckoldry), and male counter-strategies that function to reduce this risk. I discuss this form of intersexual conflict in the second section of this chapter. In the first section, I discuss paternal investment and its evolution, as well as women's strategies to secure this investment from one or several men.

Paternal investment

In the first part of this section, I briefly describe the cross-species conditions that favor the evolution of paternal investment, and in the second part I elaborate on a model of the social ecology in which human paternal investment may have evolved (see Geary, 2006; Geary & Flinn, 2001). In the final part, I describe cuckoldry risks in extant human populations and relate these to the social ecology described in the second part of the section.

EVOLUTION

Paternal investment is found in many species of bird and fish, and in some species of insect and mammal (Clutton-Brock, 1989; Perrone & Zaret, 1979; Thornhill, 1976; Wolf, Ketterson, & Nolan, 1988). Across species, paternal investment involves cost–benefit trade-offs; specifically, this investment is typically associated with a higher degree of paternity certainty and improved offspring survival and later reproductive prospects, but at a cost of lost mating opportunity (Birkhead & Møller, 1996; Møller & Cuervo, 2000; Perrone & Zaret, 1979; Trivers, 1972). The trade-offs are a reflection of the evolutionary history of the species and of the current social and ecological conditions in which the individual is situated. Evolutionary history is important because it determines whether paternal investment is *obligate* or *facultatively* expressed (Arnold & Owens, 2002; Clutton-Brock, 1991; Fishman, Stone, & Lotem, 2003). Obligate investment means that male care is necessary for the survival of his offspring and thus selection will favor males who always invest in offspring. One potential long-term result is that males will show high levels of paternal investment, independent of proximate conditions (Westneat & Sherman, 1993). Moreover, given the costs of abandonment by the male, females rarely cuckold in these species (Birkhead & Møller, 1996).

Paternal investment in humans and many other species is facultatively expressed: it often benefits offspring but is not always necessary for their survival and thus the quantity and quality of human paternal investment often varies with proximate conditions (Geary, 2000; Westneat & Sherman, 1993). The facultative expression of male parenting reflects the just-noted cost–benefit trade-offs as these relate to the current social and ecological contexts in which the male is situated. The degree of male investment (1) increases with increases in the likelihood that investment will be provided to his own offspring (i.e. paternity certainty), (2) increases when investment increases the survival and later reproductive prospects of offspring, and (3) decreases when there are opportunities to mate with multiple females. Equally important, the conditional benefits of paternal investment in these species results in simultaneous cost–benefit trade-offs in females. Sometimes it is in the females' best interest (e.g. when paired with an unhealthy male) to cuckold their partner and mate with higher-quality males (Birkhead & Møller, 1996). It is in these species – which includes humans – that prolonged intersexual conflict is predicted to evolve.

HUMAN PATERNAL INVESTMENT

Humans are among the 5% of mammalian species in which males invest in the well-being of their offspring, although as noted this investment is facultatively expressed (Geary, 2000). Here I outline the potential evolutionary history of human paternal investment and then the conditions associated with the facultative expression of this investment. The combination provides a context for understanding the issues of paternity certainty, mate guarding, and cuckoldry, as described in the final part of this section.

Evolutionary history

The evolution of human paternal behavior has almost certainly been influenced by the same cost–benefit trade-offs associated with paternal investment in other species. The evolutionary history of these trade-offs is never certain, but can be guided by comparative analyses of evolutionarily related species. For humans, the most appropriate comparisons would involve other species of *Homo* and australopithecines, but these are all extinct. Thus, a common approach is to use patterns in the two species most closely related to humans: chimpanzees (*Pan troglodytes*) and bonobos (*Pan paniscus*). Because males show little or no paternal investment in these species, it is not clear that they are appropriate comparison species (see also Geary & Flinn, 2001). If our ancestors were like chimpanzees or bonobos, multiple changes in male (e.g. increase in parenting) and female (e.g. emergence of concealed ovulation) reproductive behavior would have had to occur to create the current human reproductive pattern. Geary and Flinn (2001) proposed that the reproductive dynamics of our ancestors might instead have been more similar to that of our distant cousin, the gorilla (*Gorilla gorilla*). This is because moving from a gorilla-like pattern to the current human pattern would require fewer evolutionary changes than would be necessary to move from a chimpanzee- or bonobo-like pattern to the modal human pattern.

The modal social organization of gorillas is often described as isolated single-male harems, which typically include one reproductive male, two to four females, and their offspring (Fossey, 1984; Stewart & Harcourt, 1987; Taylor, 1997). However, there is considerable variation in this social structure, even in the most isolated groups of mountain gorillas (*G. gorilla beringei*); Robbins (1999) found that 40% of these groups included several often related males (e.g. brothers or father–sons). Encounters between groups are rare but when they do occur they result in intense and physical male–male competition over females, and male mate guarding of the females in their group (e.g. gathering them together and placing themselves between the females and the intruding male or males).

Groups of lowland gorillas (*G. gorilla gorilla*) also maintain a harem structure, but in contrast to mountain gorillas they are less socially isolated. Several families will occupy the same geographical region and encounters between groups are often friendly, especially among the males (Bradley *et al.*, 2004). Bradley *et al.*'s DNA fingerprinting of male and female relatedness among these families indicates that males tend to be organized as clusters of kin, whereas females tend to be unrelated to other group members. This patrilocal social structure is similar to that found in chimpanzees (*P. troglodytes*; Goodall, 1986) and humans in traditional societies today (e.g. Pasternak, Ember, & Ember, 1997); that is, males tend to stay in or near the territory of their birth group whereas females tend to emigrate from their birth group to the group of their mate. The kinship organization of male lowland gorillas provides a ready explanation for the lower levels of male–male competition in comparison with that found with mountain gorillas.

In any case, the dynamics that emerge within families of lowland gorillas is similar to that found in human families. Unlike the unrestricted mating of female chimpanzees (during estrous) or bonobos, and a corresponding low level of paternity certainty for conspecific chimpanzees (de Waal & Lanting, 1997; Goodall, 1986), adult male and female gorillas often form long-term social relationships. DNA fingerprinting indicates that male gorillas show high levels of paternity certainty (>95%; Bradley *et al.*, 2004), and behavioral observation has revealed low levels of mate guarding (e.g. compared to chimpanzees) and high levels of affiliation with their offspring. "Associated males hold, cuddle, nuzzle, examine, and groom infants, and infants turn to these males in times of distress" (Whitten, 1987, p. 346). Unlike female chimpanzees and bonobos, female gorillas do not typically have conspicuous sexual swellings and primarily solicit copulations behaviorally (Stewart & Harcourt, 1987). The gorilla-like pattern of female sexual solicitation is more similar to the current human pattern (e.g. concealed ovulation) than is the pattern of female solicitation in chimpanzees or bonobos.

The primary evolutionary change needed to move from a single-male harem to the multi-male, multi-female communities found with humans is the formation of male kin-based coalitions. Bradley *et al.*'s (2004) findings indicate that, at least for lowland gorillas, the male-kinship structure is very close to that currently found with humans. The primary difference is the degree of cooperation among adult males as related to coalitional competition with other groups of male kin. Such coalitions could easily arise from the social structure described by Bradley *et al.* (2004). The formation of more closely knit male kinship coalitions would create greater proximity of males and through this the creation of multi-male, multi-female communities. Indeed, if gorilla families were placed in closer proximity and if male-kinship bonds were strengthened, the common structure of human families, including polygynous families, in traditional societies would be formed. If this model is close to being correct, then paternal investment has a long evolutionary history in humans, as suggested by Lovejoy (1981), as do long-term male–female reproductive relationships.

The formation of tightly knit patrilocal communities, however, results in several additional and important changes in social organization. The pressure to maintain a male coalition creates pressures that facilitate within-group cooperation and this in turn is likely to result in reduced polygyny and a higher percentage of males that reproduce (see also Geary & Flinn, 2002). The cost of less reproductive skew among males is that more lower-quality males enter the reproductive pool and thus more females are paired with these males. The existence of multi-male groups and the pairing of lower-quality males results in greater potential for extra-pair relationships and great potential benefits to females if they cuckold lower-quality social partners. The resulting predictions are that paternity certainty is lower than that found in lowland gorillas, but much higher than that found in chimpanzees or bonobos. Paternity certainty should be associated with coevolving mechanisms, such as pair bonding, that maintain the male–female reproductive relationship in multi-male, multi-female communities (MacDonald, 1992). There will also be sex-specific adaptations that, under some conditions (e.g. when paired with an unhealthy mate), will enable the seeking of extra-pair relationships and corresponding counter strategies that will disrupt these relationships. The focus of this chapter is on cuckoldry and male mate guarding, but these are not the only adaptations predicted to evolve in this social ecology.

Proximate expression

If the above scenario is correct, then male parenting, long-term female–male relationships, and a family structure following the gorilla-like pattern were in place before the emergence of our australopithecine ancestors, as proposed originally by Lovejoy (1981). Even with a long evolutionary history, paternal investment would not have been maintained during human evolution, if it did not confer reproductive benefits to males. The benefits would occur in the context of the same trade-offs described in the previous section; specifically, as related to paternity certainty and offspring survival and later reproductive prospects, as these benefits are balanced against lost mating opportunities. These trade-offs are indeed expressed in many human populations with the facultative expression of men's parenting.

The details of the family-dynamic, cultural, and other correlates of men's paternal investment are described elsewhere (Flinn & Low, 1986; Geary, 2000, 2006). The point here is that in many contexts men's provisioning and other forms of parental investment can substantively reduce children's morbidity and mortality risks (Adler et al., 1994; Hill & Hurtado, 1996; Morrison, Kirshner, & Molho, 1977; Richner, Christe, & Oppliger, 1995; Schultz, 1991) and enhance their social-competitive competencies (Amato, 1998; Kaplan et al., 1995). In many contexts the latter can influence later reproductive prospects (Irons, 1979). These relations have been considerably muted during the past century in industrialized societies, but are nonetheless evident in traditional societies today (e.g. Hill & Hurtado, 1996; United Nations, 1985) and prior to industrialization in modern societies (e.g. Morrison et al., 1977) and thus were almost certain to have been crucial during human evolutionary history.

CUCKOLDRY RISKS

The benefits of paternal investment and the coevolution of this investment with paternity certainty result in potential reproductive trade-offs for women. Because men vary in quality (e.g. health; Shackelford & Larsen, 1997) and are readily available in multi-male, multi-female communities, women have the opportunity to cuckold their social partner and can sometimes benefit from doing so. As with other species (e.g. Birkhead & Møller, 1996), men are predicted to and do reduce their levels of parental investment when they are not investing in their own children or suspect they have been cuckolded by their partner (e.g. Daly & Wilson, 1985; Daly, Wilson, & Weghorst, 1982; Flinn, 1992). As a result, women must balance the costs of reduced paternal investment or male retaliation against the benefits of cuckoldry; that is, having their children sired by a more fit man while having their social partner assist in the rearing of these children.

Although definitive conclusions cannot be reached at this time, it appears that men may be cuckolded about 10% of the time, on average (Bellis & Baker, 1990; Flinn, 1988; Gaulin, McBurney, & Brakeman-Wartell, 1997; McBurney *et al.*, 2002). The issues are complex, however, as the rate varies significantly across cultural settings and socioeconomic status. Sasse *et al.* (1994) reported that nonpaternity rates were 1% in Switzerland, but others have reported rates greater than 20% in settings of low socioeconomic status (Cerda-Flores *et al.*, 1999; Potthoff & Whittinghill, 1965). It is also possible that some of these men are aware of the nonpaternity of the children they are raising, and thus have not been technically cuckolded.

In any case, human paternity certainty is greater than 90% and considerably higher in some populations. This level of paternity certainty is much closer to that found in gorillas (>95%) than in chimpanzees and bonobos where there is little certainty of paternity given that females have multiple mates. The level of paternity certainty in human populations is what would be predicted on the basis of the gorilla-like model. More precisely, the formation of male coalitions and the resultant emergence of multi-male, multi-female communities, along with an increase in the number of lower-quality males entering the reproductive pool, will increase the opportunity for and the benefits of cuckoldry, but the pre-existing social structure would remain largely intact. The structure, as noted, would include polygynous and monogamous families with long-term male–female relationships, moderate to high levels of paternal investment, and high levels of paternity certainty. The formation of larger communities would result in increased male intrasexual competition but muted by kinship ties, and increased intersexual conflict. The latter would result in a drift, so to speak, from higher to lower levels of paternity certainty due to cuckoldry, but this drift would be countered by risk of lost paternal investment and male retaliation.

Intersexual conflict

As noted earlier, intersexual cooperation occurs when the reproductive interests of the male and female overlap and intersexual conflict arises as individuals attempt to secure more reproductive investment from other individuals than is in the latter's best interest, as often happens with parent–offspring relationships (Haig, 1993; Trivers, 1974). Among unrelated adults the most extreme forms of intersexual conflict are rape (Thornhill & Palmer, 2000) and cuckoldry. My focus is on women's and men's behavioral and psychological adaptations that may have evolved in the context of intersexual conflict and cuckoldry. The potential for this specific form of intersexual conflict arises when women have the opportunity to develop relationships with extra-pair men and under conditions when it is in women's best interest to develop potentially reproductive relationships with one or more of these men. In the first part of this section I present an overview of the conditions most clearly associated with women developing relationships with more than one man, and in the second part I explore the social and psychological aspects of the dynamics of the relationships that are most likely to result in cuckoldry.

SERIAL MONOGAMY AND POLYANDRY

In theory, the reproductive best interest of most women can be achieved by marriage to a socially dominant and wealthy man who is able and willing to invest his high reproductive potential in her children, but this is not always achievable (Buss, 1989; Gangestad & Simpson, 2000; Geary, 1998). This appears to be particularly true in contexts populated by many low-status men who do not have the material resources to support a family. As one potential adaptation to these conditions, women might develop a successive series of relationships with a number of these men or several simultaneously, each of whom provides some investment during the course of the relationship (Buss & Schmitt, 1993; Campbell, 2002; Greiling & Buss, 2000). These women are practicing serial monogamy and sometimes polyandry, and they and their children are often healthier than women living in the same contexts but who are monogamously married to low-status men with few resources (Lancaster, 1989).

In several South American Indian societies, such as the Ache and Barí, women will engage in sexual relations with men who are not their social partner, especially after becoming pregnant (Beckerman *et al.*, 1998; Hill & Hurtado, 1996). By tradition, these men are called secondary fathers and are socially obligated to provide material resources and social protection to the woman's child, although not all of them do so. The result seems to be a confusion of paternity such that both primary and secondary fathers invest in the child. The mortality rate of Ache children with one secondary father is about half that of children with no secondary father or two or more secondary fathers (Hill & Hurtado, 1996). With more than one secondary father, paternity is too uncertain and thus these men do not invest in the child. The benefit of a secondary father cannot be attributed to qualities of the mother, as Beckerman *et al.* (1998) found that 80% of Barí children with a secondary father survived to adulthood, as compared with 61% of their siblings without a secondary father. Under these conditions, the reproductive benefits of multiple sexual relationships are clear.

In the Ache and Barí, the primary and secondary fathers of children are known to the wider social group, and in situations in which women practice serial monogamy the nonpaternity of children from previous relationships is clear. In other words these are situations in which many of the men providing some form of investment to children – presumably to maintain a sexual relationship with the children's mother – are explicitly aware that they are not the biological father. In many other contexts, women's extra-pair sex is duplicitous and often involves cuckoldry of their social partners. In these situations, women must maintain two explicit heterosexual relationships, resulting in implicit intrasexual competition between the associated males. If this type of dynamic has occurred repeatedly during human evolutionary history then there are likely to have evolved social and psychological biases that increase the prospects that women can successfully deceive cuckolded males into providing paternal investment, and coevolving male counter strategies (Daly & Wilson, 1982).

WOMEN'S CUCKOLDRY STRATEGIES

In several large but unrepresentative samples, 20–25% of adult women reported having had at least one extra-pair sexual relationship during their marriage (Essock-Vitale & McGuire, 1988; Glass & Wright, 1992), and another 25% reported developing an intimate but not sexual (at that point) extra-pair relationship (Glass & Wright, 1992). Using a nationally representative sample in the USA, Wiederman (1997) found that 12% of adult women reported at least one extra-pair sexual relationship during their marriage, and about 2% reported such a relationship during the past 12 months; Treas and Giesen (2000) found similar percentages for another nationally representative sample. These may be underestimates, given that people are reluctant to admit to extra-pair relationships. In any case, the results indicate that some women develop simultaneous and multiple opposite-sex relationships, many of which become sexual and are unknown to their social partner. The studies reviewed in the Cuckoldry risks section make it clear that some of these extra-pair relationships result in pregnancy, and appear to occur with much higher frequency for women paired with lower-quality men, as is found in other species.

The dynamics of these extra-pair relationships are likely to involve a mix of implicit (i.e. unconscious) and explicit (i.e. conscious) psychological processes (e.g. attention to symmetric facial features) and social strategies. In fact, the finding that attraction to extra-pair partners is influenced by hormonal fluctuations points to the importance of implicit mechanisms. In particular, women, as a group, show systematic changes in sexual fantasy and attractiveness to extra-pair men, among other sex-related traits, around the time of ovulation (Bellis & Baker, 1990; Gangestad & Thornhill 1998; Gangestad, Thornhill, & Garver, 2002; Geary et al., 2001; Macrae et al., 2002; Penton-Voak et al., 1999; Penton-Voak & Perrett, 2000; Thornhill & Gangestad, 1999). Women are not only more likely to fantasize about (Gangestad et al., 2002) and sometimes engage in (Bellis & Baker, 1990) an affair during this time, they are also more sensitive to and attracted by male pheromones. Gangestad and Thornhill (1998) and Thornhill and Gangestad, (1999) found that the scent of facially symmetric and thus physically attractive men was rated as more attractive and sexy than was the scent of less symmetric men, but only during this fertile time frame. Penton-Voak and colleagues found that women rate masculine faces, those with a more prominent jaw, as especially attractive around the time of ovulation (Penton-Voak et al., 1999; Penton-Voak & Perrett, 2000). Scent, facial symmetry, and a masculine jaw bone may, in turn, be proximate cues to the man's genetic fitness and social dominance (Shackelford & Larsen, 1997).

The emerging picture is one in which women appear to have an evolved sensitivity to the proximate cues of men's fitness, a sensitivity that largely operates automatically and implicitly and peaks around the time women ovulate. The implicit operation of these mechanisms enables women to assess the fitness of potential extra-pair partners without a full awareness that they are doing so. In this way, women are psychologically and socially attentive to the relationship with their primary partner and most of the time have no explicit motive to cuckold this partner. If their social partners monitor for indications of attraction to extra-pair men, which they often do (see below), then these cues are only emitted during a short time frame. Moreover, given that attraction to a potential extra-pair partner is influenced by hormonal mechanisms, often combined with some level of pre-existing and non-sexual emotional intimacy with the extra-pair male (Banfield & McCabe, 2001), many of these women may have no intention of an extra-pair sexual relationship before it is initiated. Under these conditions, the dynamics of cuckoldry may involve some level of self deception on women's part, a mechanism that facilitates their ability to keep the extra-pair relationship hidden from their social partners.

MEN'S ANTI-CUCKOLDRY STRATEGIES

As with women, men's anti-cuckoldry biases almost certainly involve a mix of implicit processes and explicit behavioral strategies that can be directed toward their mates, toward potential rivals, and toward the evaluation of the likely paternity of children born to their partners (Buss, 2002; Schmitt & Buss,

2001; Symons, 1979). I provide a brief overview of some of the better studied of these mechanisms in the following sections; specifically, relationship jealousy, mate guarding, and use of paternity cues, respectively.

Relationship jealousy

Jealousy is an affective experience that is triggered by risks to a central relationship, especially relationships with parents and mates. The most salient trigger for men's jealousy in adulthood is predicted to be real or imagined threats to paternity certainty; that is, a risk of partner infidelity (Buss *et al.*, 1992; Daly & Wilson, 1982; Symons, 1979). The psychological manifestation is sexual jealousy, which has a near-universal influence on the dynamics of men's and women's relationships (e.g. Buss *et al.*, 2000; Geary *et al.*, 1995), although the evolutionary interpretation of these dynamics and associated sex differences have been questioned (e.g. Harris, 2000, 2003). In keeping with predictions derived from evolutionary theory, Sagarin and colleagues (2003) provided evidence that men's jealousy is especially likely to be triggered by paternity threats and not to other aspects of their mates' relationships with other people. Here, it was found that men were distressed by the prospect of their partner having an extra-pair sexual relationship with another man and thus risking pregnancy, but were not distressed by the prospect of their partner having a sexual relationship with a woman and thus not risking pregnancy (Sagarin *et al.*, 2003).

When triggered, jealousy often results in a variety of behavioral responses, including male-on-female aggression (Daly *et al.*, 1982; Daly & Wilson, 1988), divorce (Betzig, 1989), the monitoring and attempted control of the social and sexual behavior of their partners (Dickemann, 1981; Flinn, 1988), enhancement of their attractiveness as a mate (Buss & Shackelford, 1997), and the monitoring of and aggression toward actual or perceived sexual rivals (Wilson & Daly, 1985). In total, these behaviors encompass tactics that function to ensure, through coercion or enticement, that their reproductive investment and that of their mate is directed toward the man's biological children.

Mate guarding

One of the more common behavioral responses to relationship jealousy is mate guarding. For men this involves reducing their partner's opportunity to mate with other men. The attendant dynamics are well illustrated by Flinn's observational study of spousal and other heterosexual relationships in a rural Trinidadian village. In this village, "13 of 79 (16.4%) offspring born ... during the period 1970–1980 were putatively fathered by males other than the mother's coresident mate. Clearly, mate guarding could have significant effects on fitness" (Flinn, 1988, p. 10). Indeed, mate guarding by men but not women was found to be a common feature of long-term relationships, although the guarding varied with the pregnancy risks of the man's partner, as found by Sagarin *et al.* (2003). Men monitored the activities less diligently and had fewer conflicts with pregnant and older wives than they did with younger and non-pregnant wives. In a related study, women reported that their partner engaged in more mate guarding during the week when the women were most likely to ovulate, the time frame when these same women reported an increase in sexual fantasy and interest in an extra-pair man (Gangestad *et al.*, 2002).

The cues that trigger men's increased mate guarding during the ovulatory time frame are not well understood, but might involve sensitivity to behavioral, physical, and perhaps olfactory changes that occur during this time (Gangestad *et al.*, 2002). One obvious change might be that women become more behaviorally attentive (e.g. flirtatious) to attractive extra-pair men, which their partners will likely monitor. There are also soft-tissue changes that result in greater symmetry in many of the traits that influence men's mate choices and through this there is an increase in women's physical attractiveness (e.g. Scutt & Manning, 1996). These changes would make the woman more attractive to their partner, as well as to extra-pair men. Although the results are not conclusive, it appears that women are not only more symmetric during the time of ovulation, they may also produce olfactory cues that signal ovulation (Singh & Bronstad, 2001; but see Thornhill & Gangestad, 1999). In the better controlled of these studies, Singh and Bronstad asked women to wear t-shirts during the time of ovulation and during a non-ovulatory phase of their menstrual cycle. Men then rated the t-shirt odors in terms of pleasantness, sexiness, and intensity. Shirts worn during the ovulatory phase were rated as more pleasant and sexy than shirts worn by the same women during the non-ovulatory phase. There were, in contrast, no phase differences for rated intensity. Men may thus be sensitive to cycle-related fertility cues.

Paternity cues

The patterns described in the previous two sections are consistent with the prediction that a variety of emotional and behavioral anti-cuckoldry biases are evident in men (Daly *et al.*, 1982). These biases largely operate prior to pregnancy. Because men parent during some or all of their children's development and often beyond this, they are also predicted to have biases that serve as post-partum paternity cues. Women, in turn, are predicted to bias the use of these same cues as a means of maintaining paternal investment, especially in situations in which paternity is ambiguous (Daly & Wilson, 1982; McLain *et al.*, 2000; Pagel, 1997). Men in particular should be sensitive to cues of resemblance to their putative offspring and invest more heavily in children they perceive as resembling themselves.

Several studies suggest that fathers do indeed bias their investment in children based on their perceived resemblance to the child (Burch & Gallup, 2000; Platek et al., 2002), although results are mixed as to whether infants and young children do in fact resemble fathers more than mothers (Christenfeld & Hill, 1995; McLain et al., 2000). Platek et al. (2002) took digital photographs of men and women and morphed them to create the face of a preschool child. Participants were then presented with a set of five such photos (their morphed photo was in half of the sets) and asked to choose the child whom they were more likely to adopt, find most attractive, most likely to spend time with and invest resources on, among other investment related items. Men were significantly more likely than women to indicate that they would invest in their self-morph. Burch and Gallup (2000) assessed men who were in mandated treatment for domestic violence and also found a child-resemblance bias for the biological children or putative biological children of these men. As the perceived resemblance to their children increased, the quality of the reported relationship with the children increased ($r = 0.60$), the severity of the injuries inflicted on the children's mothers during the domestic violence decreased ($r = -0.31$), and indicators of spousal commitment improved.

In species in which cuckoldry occurs on a regular basis and in which paternal investment is important and contingent on perceived biological relatedness to offspring, as with humans, it is in the females' and the offspring's best interest to confuse paternity (Pagel, 1997). Daly and Wilson (1982) found that in videos of spontaneous interactions in maternity wards in the USA, mothers stated that the newborn resembled the father more than the newborn resembled her but fathers were more skeptical of this resemblance. Follow-up studies confirmed the pattern in Canada and Mexico and suggest that it extends to maternal kin as well as the mother (Daly & Wilson, 1982; McLain et al., 2000; Regalski & Gaulin, 1993). Men are thus biased to invest in children whom they perceive as resembling themselves, and women and their kin are biased such that they are much more likely to provide social cues suggesting greater paternal than maternal resemblance to children; in other words, women and their kin attempt either implicitly or explicitly to manipulate social information in ways that would result in increased paternal investment.

Men may also assess the likelihood of their partner seeking an extra-pair relationship based on their partner's personality, and on the overall quality of the relationship. Men married to women who are conscientious and emotionally stable report higher levels of marital satisfaction and lower perceived risk of partner infidelity (Shackelford & Buss, 2000). When it comes to an actual infidelity, women who regularly participate in religious activities are less likely to engage in an extra-pair relationship, whereas women who are unhappy in their current relationships, have been previously divorced, and earn more money (and thus less dependent on spousal support) are more likely to engage in such a relationship, although cause and effect are difficult to separate (Atkins, Baucom, & Jacobson, 2001; Previti & Amato, 2004). In any case, a 17-year longitudinal study of family functioning revealed that beliefs about the inherent stability of the current relationship were predictive of later infidelities. In other words, individuals who believed the relationship was unstable and couples who discussed dissolution of the relationship were more likely to experience an infidelity by one or both of them in subsequent years. It follows that men may be more prone to suspect infidelity or mate guard partners who express a dissatisfaction with the relationship or who believe that relationships are generally unstable.

Sperm competition

Cuckoldry, of course, occurs when women maintain more than one sexual relationship and engage in sexual intercourse with both partners within roughly a 5-day time frame (Baker & Bellis, 1995). Under these conditions, the women's partners are implicitly competing through sperm competition. In modern populations, it is clear that some women do engage in multiple and simultaneous sexual relationships and thus create conditions that could promote sperm competition. Bellis and Baker (1990) found that when women initiated an infidelity it often occurred around the time of ovulation. For this sample, 7% of the copulations during the time of ovulation were with an extra-pair man, and these relationships were less likely to involve the use of contraceptives than were copulations with their social partner. In a national probability survey of the sexual behavior of adults in Britain, Johnson *et al.* (2001) found that 15% of the 16–24-year-old women and 8% of the 25–34-year-old women reported concurrent sexual relationships during the past year.

Although these patterns and other evidence support the hypothesis that sperm competition may have occurred during human evolutionary history (Baker & Bellis, 1993; Pound, 2002; Shackelford *et al.*, 2002), the extent of such competition and its importance in shaping human reproductive behavior is debated (e.g. Nicholls, 2002). If human paternal investment evolved in a social context that is closely related to a gorilla-like social structure embedded in multi-male, multi-female communities then sperm competition would have been considerably less important than if our evolutionary history was closer to that of the chimpanzee- or bonobo-like reproductive pattern.

Conclusion

Paternal investment and cuckoldry are necessarily linked, proximately and ultimately, and the attendant dynamics reflect a balance of cost–benefit trade-offs for both males and females (Clutton-Brock, 1991; Fishman *et al.*, 2003). Paternal investment has evolved for species in which such investment is likely to be directed toward biological offspring (i.e. high paternity certainty), is associated with improved survival and later reproductive prospects for these offspring, and does not substantively reduce mating opportunity. When paternal investment is necessary for offspring survival it is obligate and nearly all males invest and very few females cuckold (Birkhead & Møller, 1996). When paternal investment is helpful but not obligate, then it is facultatively expressed. This contingent expression of male parenting is influenced by the local ecology and social group in which the male is situated and the associated trade-offs between the male's mating opportunities, paternity certainty, and the benefits accrued by his offspring. These same conditions create the potential for females to risk cuckolding their social partners. The risks are male abandonment and retaliation and the primary benefits are having offspring sired by a more fit male – their offspring have higher survival rates – while securing investment for her offspring from the social partner.

The evolutionary history of human paternal investment is not fully understood, although it is clear that it is facultatively expressed in most ecologies and social groups (Flinn & Low, 1986) and almost certainly expressed in accordance with the same cost–benefit trade-offs found in other species (Geary, 2000, 2006). Our understanding and interpretation of these proximate patterns will be influenced by our understanding of the contexts in which human paternal investment evolved (Lovejoy, 1981). If the evolutionary history is similar to that found in extant groups of chimpanzees and bonobos, where there is little paternal investment or paternity certainty, then human paternal investment, long-term male–female relationships, and relatively high levels of paternity certainty are recently evolved traits. In contrast, if male parenting evolved in a context similar to that found in extant lowland gorillas, then these traits have a long evolutionary history. In fact, reductions in paternity certainty and cuckoldry would be the more recent evolutionary changes; these would coincide with the formation of close-knit male kin groups and the formation of large multi-male, multi-female communities.

If the gorilla-like model is correct, then men are biased as a group to invest in the well-being of their children, though still basically polygynous in their mate preferences (as with gorillas), and both men and women are biased to develop long-term (though not necessarily exclusive) reproductive relationships. From this perspective, cuckoldry risks are predicted to be greater than but still close to those found in extant gorillas, as they are. Although we await definitive results, cuckoldry rates across human populations appear to be about 10% for humans as compared with less than 5% in gorillas (Bradley *et al.*, 2004). As is found in other species with the facultative expression of paternal investment, women appear to be much more likely to cuckold lower-quality social partners or partners with whom a long-term relationship seems unlikely than they are to cuckold other men. The gist of this model is that the facultative expression of men's parenting may be closer to our evolutionary history and women's polyandrous mating and cuckoldry further from our evolutionary history then implied by evolutionary models that start from chimpanzee- or bonobo-like reproductive behavior.

References

Adler, N. E., Boyce, T., Chesney, M. A., *et al.* (1994). Socioeconomic status and health: the challenge of the gradient. *American Psychologist*, **49**, 15–24.

Alexander, R. D. (1987). *The Biology of Moral Systems*. Hawthorne, NY: Aldine de Gruyter.

Amato, P. R. (1998). More than money? Men's contributions to their children's lives. In A. Booth and A. C. Crouter, eds., *Men in Families: When Do They Get Involved? What Difference Does it Make?* Mahwah, NJ: Erlbaum, pp. 241–78.

Andersson, M. (1994). *Sexual Selection*. Princeton, NJ: Princeton University Press.

Arnold, K. E. and Owens, I. P. F. (2002). Extra-pair paternity and egg dumping in birds: life history, parental care and the risk of retaliation. *Proceedings of the Royal Society of London B*, **269**, 1263–9.

Atkins, D. C., Baucom, D. H., and Jacobson, N. S. (2001). Understanding infidelity: correlates in a national random sample. *Journal of Family Psychology*, **15**, 735–49.

Baker, R. R. and Bellis, M. A. (1993). Human sperm competition: ejaculate adjustment by males and the function of masturbation. *Animal Behaviour*, **46**, 861–85.

Baker, R. R. and Bellis, M. A. (1995). *Human Sperm Competition: Copulation, Masturbation, and Infidelity*. London: Chapman and Hall.

Banfield, S. and McCabe, M. P. (2001). Extra relationship involvement among women: are they different from men? *Archives of Sexual Behavior*, **30**, 119–42.

Beckerman, S., Lizarralde, R., Ballew, C., *et al.* (1998). The Barí partible paternity project: preliminary results. *Current Anthropology*, **39**, 164–7.

Bellis, M. A. and Baker, R. R. (1990). Do females promote sperm competition? Data for humans. *Animal Behaviour*, **40**, 997–9.

Betzig, L. (1989). Causes of conjugal dissolution: a cross-cultural study. *Current Anthropology*, **30**, 654–76.

Birkhead, T. R. and Møller, A. P. (1996). Monogamy and sperm competition in birds. In J. M. Black, ed., *Partnerships in Birds: The Study of Monogamy*. New York: Oxford University Press, pp. 323–43.

Bradley, B. J., Doran-Sheehy, D. M., Lukas, D., Boesch, C., and Vigilant, L. (2004). Dispersed male networks in Western gorillas. *Current Biology*, **14**, 510–13.

Burch, R. L. and Gallup, Jr., G. G. (2000). Perceptions of paternal resemblance predict family violence. *Evolution and Human Behavior*, **21**, 429–35.

Buss, D. M. (1989). Sex differences in human mate preferences: evolutionary hypothesis tested in 37 cultures. *Behavioral and Brain Sciences*, **12**, 1–49.

Buss, D. M. (2002). Human mate guarding. *Neuroendocrinology Letters*, **23** (Suppl. 4), 23–9.

Buss, D. M. and Schmitt, D. P. (1993). Sexual strategies theory: an evolutionary perspective on human mating. *Psychological Review*, **100**, 204–32.

Buss, D. M. and Shackelford, T. K. (1997). From vigilance to violence: mate retention and tactics in married couples. *Journal of Personality and Social Psychology*, **72**, 346–61.

Buss, D. M., Larsen, R. J., Westen, D., and Semmelroth, J. (1992). Sex differences in jealousy: evolution, physiology, and psychology. *Psychological Science*, **3**, 251–5.

Buss, D. M., Shackelford, T. K., Choe, J., Buunk, B. P., and Dijkstra, P. (2000). Distress about mating rivals. *Personal Relationships*, **7**, 235–43.

Campbell, A. (2002). *A Mind of Her Own: The Evolutionary Psychology of Women*. New York: Oxford University Press.

Cerda-Flores, R. M., Barton, S. A., Marty-Gonzalez, L. F., Rivas, F., and Chakraborty, R. (1999). Estimation of nonpaternity in the Mexican population of Nuevo Leon: a validation study with blood group markers. *American Journal of Physical Anthropology*, **109**, 281–93.

Christenfeld, N. J. S. and Hill, E. A. (1995). Whose baby are you? *Nature*, **378**, 669.

Clutton-Brock, T. H. (1989). Mammalian mating systems. *Proceedings of the Royal Society of London B*, **236**, 339–72.

Clutton-Brock, T. H. (1991). *The Evolution of Parental Care*. Princeton, NJ: Princeton University Press.

Daly, M. and Wilson, M. (1982). Whom are newborn babies said to resemble? *Ethology and Sociobiology*, **3**, 69–78.

Daly, M. and Wilson, M. I. (1985). Child abuse and other risks of not living with both parents. *Ethology and Sociobiology*, **6**, 155–76.

Daly, M. and Wilson, M. (1988). *Homicide*. New York: Aldine de Gruyter.

Daly, M., Wilson, M., and Weghorst, S. J. (1982). Male sexual jealousy. *Ethology and Sociobiology*, **3**, 11–27.

Darwin, C. (1871). *The Descent of Man, and Selection in Relation to Sex*. London: John Murray.

de Waal, F. and Lanting, F. (1997). *Bonobo: The Forgotten Ape*. Berkeley, CA: University of California Press.

Dickemann, M. (1981). Paternal confidence and dowry competition: a biocultural analysis of purdah. In R. D. Alexander and D. W. Tinkle, eds., *Natural Selection and Social Behavior*. New York: Chiron Press, pp. 417–38.

Essock-Vitale, S. M. and McGuire, M. T. (1988). What 70 million years hath wrought: sexual histories and reproductive success of a random sample of American women. In L. Betzig, M. Borgerhoff Mulder, and P. Turke, eds., *Human Reproductive Behaviour: a Darwinian Perspective*. Cambridge: Cambridge University Press, pp. 221–35.

Fishman, M. A., Stone, L., and Lotem, A. (2003). Fertility assurance through extrapair fertilizations and male paternity defense. *Journal of Theoretical Biology*, **221**, 103–14.

Flinn, M. V. (1988). Mate guarding in a Caribbean village. *Ethology and Sociobiology*, **9**, 1–28.

Flinn, M. V. (1992). Paternal care in a Caribbean village. In B. S. Hewlett, ed., *Father-Child Relations: Cultural and Biosocial Contexts*. New York: Aldine de Gruyter, pp. 57–84.

Flinn, M. V. and Low, B. S. (1986). Resource distribution, social competition, and mating patterns in human societies. In D. I. Rubenstein and R. W. Wrangham, eds., *Ecological Aspects of Social Evolution: Birds and Mammals*. Princeton, NJ: Princeton University Press, pp. 217–43.

Fossey, D. (1984). *Gorillas in the Mist*. Boston, MA: Houghton Mifflin Co.

Gangestad, S. W. and Simpson, J. A. (2000). The evolution of human mating: trade-offs and strategic pluralism. *Behavioral and Brain Sciences*, **23**, 573–644.

Gangestad, S. W. and Thornhill, R. (1998). Menstrual cycle variation in women's preferences for the scent of symmetrical men. *Proceedings of the Royal Society of London B*, **265**, 927–33.

Gangestad, S. W., Thornhill, R., and Garver, C. E. (2002). Changes in women's sexual interests and their partner's mate retention tactics across the menstrual cycle: evidence for shifting conflicts of interest. *Proceedings of the Royal Society of London B*, **269**, 975–82.

Gaulin, S. J. C., McBurney, D. H., and Brakeman-Wartell, S. L. (1997). Matrilateral biases in the investment of aunts and uncles: a consequence and measure of paternity uncertainty. *Human Nature*, **8**, 139–51.

Geary, D. C. (1998). *Male, Female: The Evolution of Human Sex Differences*. Washington, DC: American Psychological Association.

Geary, D. C. (2000). Evolution and proximate expression of human paternal investment. *Psychological Bulletin*, **126**, 55–77.

Geary, D. C. (2002). Sexual selection and human life history. In R. Kail, ed., *Advances in child development and behavior* (Vol 30). San Diego, CA: Academic Press, pp. 41–101.

Geary, D. C. (2006). Evolution of paternal investment. In D. M. Buss, ed., *The Evolutionary Psychology Handbook*. Hoboken, NJ: John Wiley & Sons, pp. 483–505.

Geary, D. C. and Flinn, M. V. (2001). Evolution of human parental behavior and the human family. *Parenting: Science and Practice*, **1**, 5–61.

Geary, D. C. and Flinn, M. V. (2002). Sex differences in behavioral and hormonal response to social threat: commentary on Taylor *et al.* (2000). *Psychological Review*, **109**, 745–50.

Geary, D. C., Rumsey, M., Bow-Thomas, C. C., and Hoard, M. K. (1995). Sexual jealousy as a facultative trait: evidence from the pattern of sex differences in adults from China and the United States. *Ethology and Sociobiology*, **16**, 355–83.

Geary, D. C., DeSoto, M. C., Hoard, M. K., Sheldon, M. S., and Cooper, L. (2001). Estrogens and relationship jealousy. *Human Nature*, **12**, 299–320.

Glass, S. P. and Wright, T. L. (1992). Justifications for extramarital relationships: the association between attitudes, behaviors, and gender. *Journal of Sex Research*, **29**, 361–87.

Goodall, J. (1986). *The Chimpanzees of Gombe: Patterns of Behavior*. Cambridge, MA: The Belknap Press.

Greiling, H. and Buss, D. M. (2000). Women's sexual strategies: the hidden dimension of extra-pair mating. *Personality and Individual Differences*, **28**, 929–63.

Harris, C. R. (2000). Psychophysiological responses to imagined infidelity: the specific innate modular view of jealousy reconsidered. *Journal of Personality and Social Psychology*, **78**, 1082–91.

Harris, C. R. (2003). A review of sex differences in sexual jealousy, including self-report data, psychophysiological responses, interpersonal violence, and morbid jealousy. *Personality and Social Psychology Review*, **7**, 102–28.

Haig, D. (1993). Genetic conflicts in human pregnancy. *Quarterly Review of Biology*, **68**, 495–532.

Hill, K. and Hurtado, A. M. (1996). *Ache Life History: The Ecology and Demography of a Foraging People*. New York: Aldine de Gruyter.

Irons, W. (1979). Cultural and biological success. In N. A. Chagnon and W. Irons, eds., *Natural Selection and Social Behavior*. North Scituate, MA: Duxbury Press, pp. 257–72.

Johnson, A. M., Mercer, C. H., Erens, B., *et al.* (2001). Sexual behaviour in Britain: partnerships, practices, and HIV risk behaviors. *Lancet*, **358**, 1835–42.

Kaplan, H. S., Lancaster, J. B., Bock, J. A., and Johnson, S. E. (1995). Does observed fertility maximize fitness among New Mexican men? A test of an optimality model and a new theory of parental investment in the embodied capital of offspring. *Human Nature*, **6**, 325–60.

Lancaster, J. B. (1989). Evolutionary and cross-cultural perspectives on single-parenthood. In R. W. Bell and N. J. Bell, eds., *Interfaces in Psychology: Sociobiology and the Social Sciences*. Lubbock, TX: Texas Tech University Press, pp. 63–72.

Lovejoy, C. O. (1981). The origin of man. *Science*, **211**, 341–50.

MacDonald, K. (1992). Warmth as a developmental construct: an evolutionary analysis. *Child Development*, **63**, 753–73.

Macrae, C. N., Alnwick, K. A., Milne, A. B., and Schloerscheidt, A. M. (2002). Person perception across the menstrual cycle: hormonal influences on social-cognitive functioning. *Psychological Science*, **13**, 532–6.

McBurney, D. H., Simon, J., Gaulin, S. J. C., and Geliebter, A. (2002). Matrilateral biases in the investment of aunts and uncles: replication in a population presumed to have high paternity certainty. *Human Nature*, **13**, 391–402.

McLain, D. K., Setters, D., Moulton, M. P., and Pratt, A. E. (2000). Ascription of resemblance of newborns by parents and nonrelatives. *Evolution and Human Behavior*, **21**, 11–23.

Møller, A. P. and Cuervo, J. J. (2000). The evolution of paternity and paternal care. *Behavioral Ecology*, **11**, 472–85.

Morrison, A. S., Kirshner, J., and Molho, A. (1977). Life cycle events in 15th century Florence: records of the Monte Delle Doti. *American Journal of Epidemiology*, **106**, 487–92.

Nicholls, H. (2002). Sperm control. *Trends in Cognitive Sciences*, **6**, 185.

Pagel, M. (1997). Desperately concealing father: a theory of parent-infant resemblance. *Animal Behaviour*, **53**, 973–81.

Pasternak, B., Ember, C. R., and Ember, M. (1997). *Sex, Gender, and Kinship: a Cross-Cultural Perspective*. Upper Saddle River, NJ: Prentice-Hall.

Penton-Voak, I. S. and Perrett, D. I. (2000). Female preference for male faces changes cyclically: further evidence. *Evolution and Human Behavior*, **21**, 39–48.

Penton-Voak, I. S., Perrett, D. I., Castles, D. L., *et al.* (1999). Menstrual cycle alters face preference. *Nature*, **399**, 741–2.

Perrone, Jr., M. and Zaret, T. M. (1979). Parental care patterns of fishes. *American Naturalist*, **113**, 351–61.

Platek, S. M., Burch, R. L., Panyavin, I. S., Wasserman, B. H., and Gallup, Jr., G. G. (2002). Reactions to children's face resemblance affects males more than females. *Evolution and Human Behavior*, **23**, 159–66.

Potthoff, R. F. and Whittinghill, M. (1965). Maximum-likelihood estimation of the proportion of nonpaternity. *American Journal of Human Genetics*, **17**, 480–94.

Pound, N. (2002). Male interest in visual cues of sperm competition risk. *Evolution and Human Behavior*, **23**, 443–66.

Previti, D. and Amato, P. R. (2004). Is infidelity a cause or a consequence of poor marital quality? *Journal of Social and Personal Relationships*, **21**, 217–30.

Regalski, J. M. and Gaulin, S. J. C. (1993). Whom are Mexican infants said to resemble? Monitoring and fostering paternal confidence in the Yucatan. *Ethology and Sociobiology*, **14**, 97–113.

Richner, H., Christe, P., and Oppliger, A. (1995). Paternal investment affects prevalence of malaria. *Proceedings of the National Academy of Sciences USA*, **92**, 1192–4.

Robbins, M. M. (1999). Male mating patterns in wild multimale mountain gorilla groups. *Animal Behaviour*, **57**, 1013–20.

Sagarin, B. J., Becker, D. V., Guadagno, R. E., Nicastle, L. D., and Millevoi, A. (2003). Sex differences (and similarities) in jealousy: the moderating influence of infidelity experience and sexual orientation of the infidelity. *Evolution and Human Behavior*, **24**, 17–23.

Sasse, G., Muller, H., Chakraborty, R., and Ott, J. (1994). Estimating the frequency of nonpaternity in Switzerland. *Human Heredity*, **44**, 337–43.

Schmitt, D. P. and Buss, D. M. (2001). Human mate poaching: tactics and temptations for infiltrating existing mateships. *Journal of Personality and Social Psychology*, **80**, 894–917.

Schultz, H. (1991). Social differences in mortality in the eighteenth century: an analysis of Berlin church registers. *International Review of Social History*, **36**, 232–48.

Scutt, D. and Manning, J. T. (1996). Symmetry and ovulation in women. *Human Reproduction*, **11**, 2477–80.

Shackelford, T. K. and Buss, D. M. (2000). Marital satisfaction and spousal cost-infliction. *Personality and Individual Differences*, **28**, 917–28.

Shackelford, T. K. and Larsen, R. J. (1997). Facial asymmetry as an indicator of psychological, emotional, and physiological distress. *Journal of Personality and Social Psychology*, **72**, 456–66.

Shackelford, T. K., LeBlanc, G. J., Weekes-Shackelford, V. A., *et al.* (2002). Psychological adaptation to human sperm competition. *Evolution and Human Behavior*, **23**, 123–38.

Singh, D. and Bronstad, P. M. (2001). Female body odour is a potential cue to ovulation. *Proceedings of the Royal Society of London B*, **268**, 797–801.

Stewart, K. J. and Harcourt, A. H. (1987). Gorillas: variation in female relationships. In B. B. Smuts, D. L. Cheney, R. M. Seyfarth, R. W. Wrangham, and T. T. Struhsaker, eds., *Primate Societies*. Chicago, IL: University of Chicago Press, pp. 155–64.

Symons, D. (1979). *The Evolution of Human Sexuality*. New York: Oxford University Press.

Taylor, A. B. (1997). Relative growth, ontogeny, and sexual dimorphism in gorilla (*Gorilla gorilla gorilla* and *G. g. beringei*): evolutionary and ecological considerations. *American Journal of Primatology*, **43**, 1–31.

Thornhill, R. (1976). Sexual selection and paternal investment in insects. *American Naturalist*, **110**, 153–63.

Thornhill, R. and Gangestad, S. W. (1999). The scent of symmetry: a human sex pheromone that signals fitness? *Evolution and Human Behavior*, **20**, 175–201.

Thornhill, R. and Palmer, C. T. (2000). *A natural history of rape: biological bases of sexual coercion*. Cambridge, MA: MIT Press.

Treas, J. and Giesen, D. (2000). Sexual infidelity among married and cohabitating Americans. *Journal of Marriage and the Family*, **62**, 48–60.

Trivers, R. L. (1972). Parental investment and sexual selection. In B. Campbell, ed., *Sexual Selection and the Descent of Man 1871–1971*. Chicago, IL: Aldine Publishing. pp. 136–79.

Trivers, R. L. (1974). Parent-offspring conflict. *American Zoologist*, **14**, 249–64.

United Nations (1985). *Socio-Economic Differentials in Child Mortality in Developing Countries*. New York: United Nations.

Westneat, D. F. and Sherman, P. W. (1993). Parentage and the evolution of parental behavior. *Behavioral Ecology*, **4**, 66–77.

Whitten, P. L. (1987). Infants and adult males. In B. B. Smuts, D. L. Cheney, R. M. Seyfarth, R. W. Wrangham, and T. T. Struhsaker, eds., *Primate Societies*. Chicago, IL: University of Chicago Press. pp. 343–57.

Wiederman, M. W. (1997). Extramarital sex: prevalence and correlates in a national survey. *Journal of Sex Research*, **34**, 167–74.

Wilson, M. and Daly, M. (1985). Competitiveness, risk taking, and violence: the young male syndrome. *Ethology and Sociobiology*, **6**, 59–73.

Wolf, L., Ketterson, E. D., and Nolan, Jr., V. (1988). Paternal influence on growth and survival of dark-eyed junco young: do parental males benefit? *Animal Behaviour*, **36**, 1601–18.

PART II MATE GUARDING

3

Evidence for adaptations for female extra-pair mating in humans: thoughts on current status and future directions

STEVEN W. GANGESTAD
University of New Mexico

Introduction

In the late 1980s and early 1990s, behavioral ecologists were first able to estimate, with good reliability, the paternity of broods in socially monogamous birds. Within a very few years, rates of paternity by social fathers had been estimated within a large number of populations. An amazing pattern emerged from these studies, one that changed fundamentally the way that behavioral ecologists viewed mating in birds. The extra-pair paternity rate – the proportion of offspring not sired by social fathers – was, on average, 10–15% (e.g. Birkhead & Møller, 1995; Petrie & Kempenaers, 1998). Though occasionally in the 1–3% range, estimated extra-pair paternity rates of at least 20% were not uncommon, and they reached 50% in a few populations. A substantial proportion of socially monogamous bird species are clearly not sexually monogamous.

From the perspective of some observers, these surprising empirical findings were the leading edge of a revolutionary "paradigm shift" in behavioral ecology currently taking place, with "the traditional concepts of the choosy, monogamous female and the coadapted gene complex increasingly giving way to the realization that sexual reproduction engenders conflicts [and] promotes polyandry ..." (Zeh & Zeh, 2001). Though Trivers' (1972) parental investment theory did not explicitly claim that females in socially monogamous species should be sexually monogamous, it did emphasize the potential reproductive benefits of male multiple mating and not the reasons why females (the sex typically investing more heavily in offspring) might multiply mate. Its arguments were based on reasoning about limitations on offspring number. A sex investing little in offspring can, potentially, greatly increase reproductive success (meaning offspring number here) by accessing a greater number of mates – the presumed reason why males in many species seek multiple mates. A sex investing heavily in offspring can produce only a limited number of offspring, and that number is not dependent on how many mates are accessed.

As Hrdy (1979) observed, while female multiple mating may not lead to benefits in terms of offspring number, it can nonetheless have benefits by enhancing *offspring quality* – the probability that offspring will survive and reproduce themselves. She emphasized the function of paternity confusion: by leading multiple males to think they could have sired her offspring, a female can reduce the probability that her offspring will be killed by a non-father. Her view remains a leading theory of female multiple mating in group-living primate societies (see, for instance, Kappeler & van Schaik, 2003).

This theory, however, appears not to apply to many bird species, in which young are not typically subject to attack by non-resident males. Bird species are also different from the typical case in which paternity confusion may operate in the sense that social mates do provide substantial material benefits to offspring, which would seem to give reason to a female to convince one particular male that he *is* the father of her offspring. The fact that social monogamy in birds so often coexists with female sexual infidelity to social mates gives rise to a fundamental question: what benefits do females in these species derive from multiple mating, benefits that presumably offset the potential loss of investment from a social mate?

Several leading theories now focus on *genetic benefits* – benefits that derive from the genes females pair their own genes with in their offspring, which affect offspring quality. There are three kinds of genetic benefit for offspring females could derive from multiple mating (for a review, see Jennions & Petrie, 2000).

INTRINSIC GOOD GENES

Over evolutionary time, natural selection sifts through available genetic variants, saving those that promote success within a species' niche and discarding others. Our genes are typically "good genes" that have passed a test of time. But some aren't. Genes mutate. Though each gene is copied correctly more than 99.99% of the time, sperm or eggs commonly contain one or more new copying errors. Because mutations typically have minor effects (much as slight impurities in a tank of gasoline subtly compromise car performance), most survive multiple generations before being eliminated. On average, we probably have several tens if not hundreds of mutations in our genomes. It was once assumed that the relentless selection against deleterious mutations would typically render their impact on variation in fitness within populations negligible. Theoretical developments beginning in the late 1980s (e.g. Rice, 1988) – proceeding independently of the findings on avian infidelity – showed this not to be the case. While selection eliminates mutations, there is an equilibrium state at which the fitness effects of mutations taken out of the population per generation are equal to the fitness effects of those entering the population. A standardized measure of genetic variation is the additive genetic coefficient of variation of CV_A, which is the square root of the additive genetic variance of a trait (in a sense, the "additive genetic standard deviation," though that term is typically not used) divided by the trait mean. (The CV_A is meaningful only for traits measured on a ratio-level scale.) The CV_A of typical traits is about 4–5 (e.g. Pomiankowski & Møller, 1995). This is true of human height, for instance: male adult height has a mean of about 70 inches and a standard deviation of about 3–3.5 inches, most of which reflects genetic variation, giving a CV_A of about 4–5. Mutation itself appears to be able to account for a CV_A in fitness of 8–20 (e.g. Charlesworth, 1990; Charlesworth & Hughes, 1998). Additionally, the world to which we must adapt isn't constant. In many ways (e.g. its gravitational fields) it is, yet in other subtle but profound ways it isn't. Pathogens constantly evolve to better thrive in our bodies and we must change merely to keep pace. Despite selection on thousands of ancestral generations to resist pathogens, we don't possess sure-fire defenses against them (e.g. Hamilton & Zuk, 1982). Empirical estimates of the CV_A of fitness are in the range of 10–30, much greater than typical morphological traits (Burt, 1995; Houle, 1992). If fitness has a CV_A of 20 – the middle of that range – individuals near the top of the distribution (two standard deviations above the mean) would have more than double the fitness of individuals near the bottom (two standard deviations below the mean). The fitness consequences of mating with one or the other could be very substantial.

COMPATIBLE GENES

Intrinsic good genes are genes that are good for all mate choosers, presumably because these genes have additive effects on fitness. Compatible genes are also good genes, but only in conjunction with the mate chooser's own genes; they enhance fitness in combination with the chooser's genes. For compatible gene effects to exist, there naturally must be genetic variation. And some of the effects of that genetic variation on fitness must rely on dominance (heterozygote superiority) or epistatic effects. Examples may be found in the major histocompatibility complex (MHC) of humans and other species. MHC genes code for cell-surface markers that the immune system uses to detect self and hence pathogens. MHC alleles are expressed codominantly. Both alleles of individuals possessing two different alleles can present foreign peptides and hence heterozygotes can potentially detect a wider range of pathogens than homozygotes. Evidence on mice indeed shows that MHC heterozygosity has an advantage in the presence of multiple strains of pathogens. In terms of mate choice, then, individuals can enhance offspring fitness by mating with an individual who possesses MHC alleles that differ from their own. Preferences for MHC-dissimilar mates detected through signatures of MHC in scent have indeed been found in a variety of species (e.g. Penn & Potts, 1999). Studies on humans have yielded mixed results (Thornhill *et al.*, 2003; Wedekind *et al.*, 1995; Wedekind & Füri, 1997). Although one might expect individuals to typically pair up with individuals with compatible MHC alleles, due to trade-offs with other characteristics and constraints posed by mate search, some females may end up with non-compatible mates, and these females could potentially benefit from extra-pair mating to obtain genetic benefits. (The evidence on MHC-disassortative mating in humans is also mixed; Hedrick & Black [1997] and Ihara *et al.* [2000], found negative results.) Zeh and Zeh (2001) discuss specific examples of extra-pair mating in females to obtain compatible genes in non-human species.

DIVERSE GENES

The offspring of a single pair are less genetically diverse than offspring of females paired with multiple males. Diverse genetic make-ups can be advantageous as a form of bet hedging in the face of changing environments. If one genetic make-up is not well-adapted to existing environments, perhaps another is. In addition, obtaining genes from different fathers can be another way to diversify disease resistance in offspring (see the discussion of compatible genes above).

Evidence in favor of each of these models of multiple mating by females exists for some species, though extra-pair mating in some bird species cannot be accounted for by any of them (see Jennions & Petrie, 2000). What about humans? Did ancestral women extra-pair mate? If so, did they evolve adaptation to obtain genetic benefits? And if so, which form(s) of genetic benefit? (For a discussion of other candidate reasons for female infidelity, see Greiling & Buss, 2000).

The extra-pair paternity rate in human populations

Data that would appear relevant to these questions are the extra-pair paternity (EPP) rates in human populations, the proportions of offspring in socially monogamous unions not sired by the social father. Estimates of the EPP rate can be made through blood or DNA testing. K. G. Anderson (unpublished work) recently compiled estimates from nearly 50 studies. Studies are of three types. First, some studies examine the EPP rate in populations of high paternity confidence. These samples are biased toward married couples in which both partners were willing to participate, such that men were probably more likely than randomly chosen men to be fathers of their putative offspring. Second, some studies specifically examine the EPP rate in men seeking paternity tests, samples undoubtedly biased toward high EPP rates relative to the general population. Third, a few studies report EPP rates from populations of unknown paternity confidence, typically because they are secondarily reported in other papers and hence their details are missing.

The median and mean EPP rates in samples of high paternity confidence are about 2–3%. In most modern populations in which EPP rates have been estimated, then, these estimates have been quite low. Two caveats must be added, however. First, as noted above, these samples are biased toward high paternity confidence. Not surprisingly, median estimates in samples of low and unknown paternity confidence are higher (30 and 17%, respectively). The actual EPP rate in populations is presumably some weighted function of these estimates, though probably typically much closer to the high-paternity- than low-paternity-confidence median (K. G. Anderson, unpublished work). Second, even within the high-paternity-confidence studies, the estimates are quite variable. Most in developed Western countries have been very low (less than 1 to about 8% [the latter study in France]). Estimates in more traditional Central and South American cultures have been higher: 3% (Mexico), 9% (Yanomano, Venezuela), and 12% (Monterrey, Mexico). The latter study found that rates varied across socioeconomic status; a low-socioeconomic-status group had an estimated rate of 20%. Not all traditional cultures have higher rates, however. The !Kung of Africa were estimated to have a rate of just 2%.

Only very tentative conclusions about typical EPP rates throughout recent human history (e.g. in the past 50 000 years) can be drawn from these data. It seems reasonable to suggest that rates have typically been less than 10% and perhaps in most cases less than 5%. It also seems reasonable to suggest that they have probably also been variable across time and place, with some populations characterized by rates of 10% or higher.

Why the EPP rate may not address questions about female adaptation for extra-pair mating

Even if reflective of ancestral rates, these relatively low average EPP rates do not directly say much about whether females possess adaptations for seeking extra-pair mating – features that have evolved because they benefited females by leading them to have sex with males other than social partners, who provided material or genetic benefits. Females don't evolve such adaptations in a vacuum. All else being equal, female extra-pair mating is clearly not in the interest of male social partners. One would therefore expect males to evolve counter-strategies to limit female extra-pair mating, which should suppress female extra-pair mating. Even when the EPP rate is fairly low, females may possess adaptations that evolved in the context of seeking extra-pair matings (and hence be part of the sum of mating adaptations females possess), which are rarely achieved because they are resisted by male social partners.

Indeed, there is reason to suspect that there may be little to resist the evolution of extra-pair mating in females in any socially monogamous species in which pairs have neighboring males *aside from* the evolution of male counter-strategies. In such species, a nearly inevitable genetic conflict of interest between the sexes arises. Not all females could possibly be mated with males who possess good genes and, hence, at least some females will have neighboring males with better genes than their partners offer. There are probably costs to having offspring sired by multiple fathers, all else equal, due to increased conflict between siblings (which, for instance, increases maternal–fetal conflict, leading to additional costs of reproduction; e.g. Haig, 1993). Extra-pair mating is common, however, suggesting that its benefits (e.g. genetic benefits) often outweigh these costs. The main deterrent to female extra-pair mating, then, is probably male resistance to it (e.g. discriminative paternal solicitude to invest in offspring of high paternity confidence, attempts to prevent or deter female social partners from engaging in extra-pair mating). Selection hence operates on each sex against the interests of the other sex; "sexually antagonistic adaptations" evolve (e.g. Rice & Holland, 1997). Females may evolve adaptations that increase their ability to obtain benefits through extra-pair mating. Males evolve counter-adaptations to reduce female partners' ability to do so. Depending on which sex evolves more effective adaptations (which may depend on ecological factors affecting the ease with which males mate-guard, the relative value of good genes, and male assistance for females, etc.), the actual extra-pair sex rate may be high (20% or more) or low (5% or less). *Even when it is low*, however, the genetic conflict of interest exists and sexually antagonistic adaptations may evolve.[1]

The question of whether women possess adaptations for extra-pair mating and, if so, what benefits led to their evolution, cannot be addressed by estimating the EPP rate itself; it must be addressed by looking directly for evidence

---

[1] The EPP rate cannot be zero and yet extra-pair mating still exert a selective pressure on a population. The point here is that the EPP rate probably not need be high to do so. Simulation modeling may provide insight into how the EPP rate may evolve with sexually antagonistic coevolution.

of those adaptations in women. Do women possess features that can reasonably be explained only in terms of selection for those features due to specific benefits garnered from extra-pair mating?

DO WOMEN POSSESS ADAPTATIONS FOR EXTRA-PAIR MATING?

The answer to this question is unknown at the present time. Here I present a brief overview of one attempt to address the question. Gangestad and Thornhill (1998) proposed to look for human adaptations that are footprints of selection forces for female extra-pair mating based on the fact that women are fertile during a brief window of their cycles. If ancestral females benefited from multiple mating to obtain genetic benefits, but at some potential cost of losing social mates or their investment, selection may have shaped preferences for indicators of those benefits to depend on fertility status: be maximal at peak fertility and less pronounced outside the fertile period. Cycle shifts should furthermore be specific to when women evaluate men as short-term sex partners (i.e. their "sexiness") rather than as long-term, investing mates (Penton-Voak *et al.*, 1999). The logic is that costs don't pay when benefits can't be reaped. The logic of a design argument for adaptation is that here is a set of features that may be expected to have evolved if females evolved to seek genetic benefits for offspring through extra-pair mating and, furthermore, may be difficult to explain as the outcome of an alternative evolutionary scenario (though see below).

Changes in female preferences across the cycle

The main indicators of genetic benefits that have been explored to date are purported indicators of intrinsic (versus compatible) good genes. To date, over a dozen studies have shown that female preferences do indeed change across the cycle. At mid cycle, women particularly prefer:

THE SCENT OF SYMMETRICAL MEN

Asymmetry on bilateral traits that are symmetrical at the population level (e.g. finger length, ear dimensions, wrist width) reflects developmental instability, perturbations due to mutations, pathogens, toxins, and other stresses. In four studies, men wore T-shirts for two nights and women rated the attractiveness of the shirts' scents (Gangestad & Thornhill, 1998; Rikowski & Grammer, 1999; Thornhill & Gangestad, 1999; Thornhill *et al.*, 2003). All found that women particularly prefer the scent of symmetrical men when fertile. Although the chemical mediating this effect has not been identified, data and theory suggest androgen-derived substances, the scent of which women evaluate more positively when fertile (e.g. Grammer, 1993; Hummel *et al.*, 1991).

MASCULINE FACES

Male and female faces differ in various ways (e.g. size of the chin, the brow ridge). Women prefer more masculine faces when they are fertile, particularly when they rate men's sexiness, not their attractiveness as long-term mates (e.g. Johnston *et al.*, 2001; Penton-Voak *et al.*, 1999).

BEHAVIORAL DISPLAYS OF SOCIAL PRESENCE AND
INTRASEXUAL COMPETITIVENESS

Gangestad *et al.* (2004) had women view videotapes of men interviewed for a potential lunch date. Men independently rated as confident and who acted toward their male competitors in condescending ways were particularly sexy to fertile women.

VOCAL MASCULINITY

At mid cycle, women find masculine (deep) voices more attractive when rating men's short-term attractiveness (Putz, 2005).

TALENT VERSUS WEALTH

Haselton and Miller (2005) found that, when faced with trade-offs between talent (e.g. creativity) and wealth, women choose talent more often when fertile, but only when evaluating men's short-term mating attractiveness.

Symmetry, masculine facial and vocal qualities, intrasexual competitiveness, and a variety of forms of talent may well have been indicators of good genes ancestrally. Not all positive traits are sexier mid cycle, however. Those more valued in long-term mates (e.g. promising material benefits) may actually be particularly preferred by non-fertile women (see Thornhill *et al.*, 2003).

Changes in female sexual attraction across the cycle

Patterns of women's sexual interests also shift across the cycle. In one study, normally ovulating women reported thoughts and feelings over the past 2 days at two times during their cycle: once when fertile (as confirmed by a luteinizing hormone surge, 1–2 days before ovulation) and once when infertile. When fertile, women reported greater sexual attraction to and fantasy about men other than primary partners – but not primary partners (Gangestad, Thornhill, & Garver, 2002; cf. Pillsworth, Haselton, & Buss, 2004; see also Bellis & Baker, 1990).

In fact, however, the hypothesis that changes evolved in the context of extra-pair mating expects a more complex pattern. On average, ancestral women could have garnered genetic benefits through extra-pair mating, but those whose primary partners had "good genes" could not. Selection thus should have shaped interest in extra-pair men mid cycle to depend on partner features; only women with men who, relatively speaking, *lack* purported indicators of genetic benefits should be particularly attracted to extra-pair men when fertile. Gangestad, Thornhill, and Garver-Apgar (2005b) tested this prediction in a replication and extension of Gangestad *et al.* (2002). Romantically involved couples participated. Again, individuals privately filled out questionnaires twice, once when the female was fertile and once during her luteal phase. Men's symmetry was measured. Once again, women reported greater attraction to extra-pair men, but not primary partners, when fertile. Effects, however, were moderated by the symmetry of women's partners. At high fertility, women with relatively asymmetrical partners were particularly attracted to extra-pair men – and less attracted to their own partners. No such effects were found during the luteal phase. Controlling for relationship satisfaction, another important predictor of women's attraction to extra-pair men, did not diminish the effect of partner symmetry.

Changes in male vigilance across the cycle

If women have been under selection to seek good genes mid cycle, men should have been under selection to take additional steps to prevent them from seeking extra-pair sex at this time. Multiple studies indicate that they do so by being more vigilant, proprietary, or monopolizing of mates' time (e.g. Gangestad *et al.*, 2002; M. G. Haselton & S. W. Gangestad, unpublished work).

There are several candidate cues of fertility status men might use. Three studies found that men find the scent of ovulating women particularly attractive (e.g. Thornhill *et al.*, 2003) and one found that men judge women's faces more attractive mid cycle. If women's interests change across the cycle, their behavior might too. Whatever the cues, women are unlikely to have been designed through selection to send them. Women do not have obvious sexual swellings mid cycle and have sex throughout the cycle, features that may well be due to selection on women to *suppress* signs of fertility status. Men, nonetheless, should be selected to detect byproducts of fertility status not fully suppressed. Consistent with this idea, Gangestad *et al.* (2002) found that enhanced male vigilance of partners mid cycle (as reported by women) was predicted by enhanced female interest in *extra-pair men*, not partners. Men may be particularly vigilant of their partners mid cycle, when their partners least want them to be.

Alternative explanations

Are these data compelling evidence for a selective history of women obtaining specific benefits through extra-pair sex, resulting in specific adaptations that evolved in the context extra-pair mating? For these data to provide truly compelling evidence, there must be no alternative explanation – that is, another evolutionary scenario that would have led to the features of women's preferences and sexual interests observed in these studies – that could account for these findings. The classic example of evidence of specific selection and function through an argument of design is the case of the vertebrate eye. Many of the details of the eye can be explained by selection for their optical properties because they allowed their owners to see and, furthermore, it is very difficult to explain them otherwise (e.g. Williams, 1992). The data on ovulatory cycle shifts, though intriguing, do not yet permit strong evidence of specific selection pressures within the context of extra-pair mating. Solid alternative explanations of findings to date, even if not yet available, may yet come forward.

One alternative theory was recently put forward by David Buller (2005). He argued that the ovulatory-cycle effects can be explained by appeal to three adaptations that did not evolve in the context of extra-pair mating: (a) the "sex drive," a desire for regular and fulfilling sex, together with efforts to satisfy that desire; (b) a peak in sexual desire during the fertile phase of the ovulatory cycle, resulting in greater female-initiated sexual activity; (c) a preference for symmetrical males, particularly when women seek partners who will satisfy sexual desires. These three adaptations could result in the empirical patterns observed as follows. When a woman is *sexually* dissatisfied in her relationship, she will be more likely to have extra-pair sex should an opportunity to do so present itself. In selecting an extra-pair partner, she will use the same criteria that she uses to choose a long-term mate but, because her desire for sexual satisfaction is heightened, she will weight factors related to sexual satisfaction, including symmetry, most highly. Women hence are expected to initiate extra-pair sex with symmetrical men more often than expected at random. Moreover, because the sex they initiate will tend to be directed toward their extra-pair partners (given that sexual dissatisfaction led to the affairs in the first place), their sex with their extra-pair partners will tend to occur when women are most likely to initiate sex – at mid cycle. Patterns of sex with and sexual interest in extra-pair partners, Buller argued, are therefore merely byproducts of adaptations that evolved in the context of female mating more generally. Hence, they constitute no evidence of female adaptation that specifically evolved in the context of extra-pair mating.

Surely, alternative explanations for ovulatory-cycle variations in women's preferences and sexual interests are welcomed. They also, however, should be subjected to the same level of close scrutiny and skepticism as the hypothesis that these variations evolved in the context of extra-pair mating. Buller's (2005) theory, in my view, leaves so many questions unanswered and findings unexplained that, even if it cannot now be dismissed as inadequate, it can be seriously questioned.

First, a crucial piece of argument in Buller's theory is that female sexual desire is enhanced mid cycle. There is indeed evidence that many women report greater sexual desire mid cycle (e.g. Regan, 1996). But, whereas some reports indicate that women furthermore are more likely to initiate sex mid cycle (e.g. Adams, Gold, & Burt, 1978; Hendricks *et al.*, 1987; Matteo & Rissman, 1984; Wilcox *et al.*, 2004), the evidence on this score is in fact very mixed, with some studies finding no effect (e.g. Bancroft *et al.*, 1983; Persky *et al.*, 1978; Schreinerss-Engel *et al.*, 1981). Indeed, evidence that women experience heightened generalized sexual desire mid cycle is mixed. Some studies that find increased attraction to men other than primary partners mid cycle have failed to find that women also report greater experience of sexual desire mid cycle (e.g. Gangestad *et al.*, 2002; M. G. Haselton & S. W. Gangestad, unpublished work). The effect of fertility status on a generalized increase in sexual desire mid cycle is probably very weak or, possibly, not truly a generalized increase in sexual desire but rather a result of increased attraction to men who exhibit particular traits (e.g. indicators of genetic benefits to offspring, as the extra-pair hypothesis proposes). As Buller's account argues that increased sexual desire mid cycle drives increased interest in extra-pair partners (in only a subset of mateships), it requires that the mid cycle increase in generalized sexual desire is substantially greater than enhanced attraction to extra-pair men mid cycle, and the available evidence casts doubt on this prediction.

Second, Buller's argument states that sexual desire evolved to be peak when women are fertile because it was designed for "reproduction." The argument seems to be that, because women can conceive only during the fertile period, they have evolved to initiate sex when fertile. This argument begs many questions. If women can conceive only when fertile, why have they evolved to seek sex (or be receptive to it) outside of the fertile phase at all? A variety of arguments for extended sexuality (sex outside of the fertile period) in humans have been proposed, and they generally concern women's ability to obtain material benefits through parental investment from males. Alexander and Noonan (1979), for instance, proposed that extended sexuality, along with concealed ovulation, enhances the pair-bond and hence evolved to facilitate biparental care in humans. Buller's theory that heightened female sexual desire occurs for the function of "reproduction" does not situate "reproduction" in any particular context and hence is not a sufficient account. At the least, if he were to acknowledge that sex outside of the cycle also evolved because it benefited female "reproduction" (albeit not directly through conception) in some way, he must address questions of why females should have evolved to weight male features differently when fertile than when outside of the fertile phase. The fact that female preferences change across the cycle mean that they have evolved to leave open the possibility that a female may be attracted to partners other than primary investing partners when mid cycle. The extra-pair mating hypothesis of ovulatory cycle shifts, of course, focuses on this feature; it argues that females initiate sex more mid cycle because they have evolved to choose sires for off-spring when mid cycle, even if those sires are different from primary investing mates. Buller's alternative explanation does not situate changes in female pre-ferences within any particular account of female reproductive strategies (or male–female relations in which reproduction occurred in contexts in which purported adaptations evolved) more generally and, therefore, is woefully incomplete at best.

Third, and relatedly, Buller does not explain why women find some desired features "sexually" attractive, whereas others are found to be good in a mate but not sexually attractive. Indeed, he appears to acknowledge that symmetry may well be found particularly sexually attractive because it was ancestrally an indicator of good genes. Ability or willingness to invest in a mate are important to mate choice, he notes, but are not found sexually attractive. But why would women have evolved to find indicators of good genes particularly *sexually* attractive, but not indicators of investment? He provides no explanation but rather takes this "fact" as a starting point. The extra-pair mating hypothesis actually does not propose that women don't find indicators of investment sexually attractive. Some may well do so, particularly in certain contexts. Hence, for instance, in the UK and Japan, women appear to find men who have feminine faces more attractive overall, purportedly because these faces advertize warmth. The face they find most attractive when mid cycle, however, is more masculine than the one they find most attractive during the luteal phase (e.g. Penton-Voak *et al.*, 1999). Buller's account is unable to explain these findings.

Fourth, Buller's argument is not able to explain some of the specific findings of ovulatory-cycle shifts. Though not explaining precisely why women find particular features sexually attractive (as just noted), at the least his theory appears to suggest that whatever women find sexually attractive in general, they will find particularly appealing mid cycle, when they can conceive. In fact, however, that's simply not the case. One exception is the shift toward favoring a more masculine face mid cycle than during the luteal phase in the UK and Japan, as just noted. Another example is discussed by Thornhill *et al.* (2003). These researchers found that women exhibit a strong preference for the scent of men who are heterozygous at MHC loci; they rate shirts worn by men hetero-zygotic at all three MHC loci as "sexier" and more "pleasant" than shirts worn by men homozygotic at one or more loci. This preference is, in fact, stronger than women's preference for the scent of symmetrical men (itself not weak). There are two reasons why women may have evolved to be relatively attracted to the scent of MHC heterozygotes. First, as discussed earlier, MHC heterozygosity may be favored in the presence of multiple strains of pathogens because it enhances disease resistance. Ancestrally, then, MHC-heterozygotic men may have been able to provide better care or a longer period of care than men homozygotic at one or more loci. Second, men heterozygotic at MHC loci produce a family of offspring that is more genetically diverse for MHC than men homozygotic at one or more MHC loci. Diversity at MHC within families may be adaptive because, should a pathogen become adapted to one offspring's make-up (by, for instance, adapting to the individuals MHC-mediated means of detecting pathogens), it's best that the pathogen not immediately be adapted to other family members' make-up. This benefit is achieved only when a female mates with a male for long-term mating and has multiple offspring with him. In producing any single offspring, a male passes on just one allele at each MHC locus; a heterozygotic mate is not necessarily more likely to produce a heterozygotic offspring than is a homozygotic mate. Both of the possible benefits of MHC heterozygosity in a mate, then, are particularly important in the context of *long-term* mating. The extra-pair mating hypothesis, then, gives no reason to expect that women will prefer the scent of MHC heterozygotic men when fertile. (Indeed, if this preference trades-off against other preferences that get more heavily weighted mid cycle – e.g. the scent of symmetrical men – it may be expected to even weaken mid cycle.) The scent is found relatively "sexy," however, and hence Buller's theory must predict that it, just like the scent of symmetry, will be particularly preferred when women are mid cycle. In fact, Thornhill *et al.* (2003) found a marginally significant trend for the scent of MHC heterozygosity to be preferred outside of the fertile phase. The difference in the pattern of cycle shifts in preference for the scent of symmetry and the shifts in preference for the scent of MHC heterozygosity was highly significant.

Fifth, Buller's account appears to be unable to account for recent findings about the role of sexual satisfaction in accounting for cycle shifts. Gangestad *et al.* (2005b), once again, found that women paired with asymmetrical men are particularly likely to experience heightened attraction to men other than partners mid cycle and experience less attraction to their partners mid cycle, relative to women paired with symmetrical men. According to Buller's account, these associations should be mediated by sexual satisfaction. Controlling for sexual satisfaction, then, should eliminate the effects. In fact, however, controlling for relationship satisfaction did not reduce the size of the effect of partner asymmetry on changes in female sexual interests at all (Gangestad *et al.*, 2005b).

Additional analyses of these data show that controlling for *sexual* satisfaction specifically also does not reduce the size of the effect. Moreover, though sexual satisfaction not surprisingly predicts sexual attraction to partners relative to attraction to other men, it did not have differential effects across the cycle in this study; Buller's theory suggests it should. Similarly, M. G. Haselton and S. W. Gangestad (unpublished work) found that women who report that they find their partner less sexually attractive (relative to their attraction as an investing mate) are particularly likely to be attracted to and flirt with men other than partners mid cycle. Controlling for a measure of sexual satisfaction in the relationship did not eliminate this effect.

In sum, Buller's alternative byproduct account for changes in female sexual interests across the cycle cannot explain a number of key findings and, furthermore, leaves a variety of important questions unanswered. Again, however, though Buller's account appears inadequate in current form, that's not to say that no alternative account can or will be proposed to explain the ovulatory-cycle shifts. The extra-pair mating account is the leading explanation of the findings to date, but is still under scrutiny.

Directions for future research

No one finding is likely to let us decide whether specific features of female mating psychology have been shaped historically within the context of extra-pair mating. Ultimately, whether there do exist such features, what those features are, and what particular benefits historically led those features to be selected will likely be decided by the slow, gradual accumulation of evidence of a variety of sorts. A few profitable lines of research come to mind.

First, most of the findings on ovulatory-cycle shifts purportedly explained by benefits acquired through extra-pair mating come from studies of college students in Western countries (largely, the USA and the UK). College students in modern Western countries are people (and partly products of selection on ancestors of modern humans) too and, hence, their behavior is undoubtedly pertinent to evaluating selectionist hypotheses. At the same time, it would be useful to know the robustness of ovulatory-cycle effects in a variety of cultural settings, including traditional, relatively stable, face-to-face societies in which individuals typically know all other members of the community. Do women experience cycle shifts in preference and patterns of attraction? Are they influenced by partner characteristics? Are there consistent patterns of change in interest in particular men and, if so, what are the features of those men? Are these effects dependent upon particular cultural features (e.g. particular levels of male proprietariness)?

Second, some researchers have attempted to appreciate the nature of female adaptations for female extra-pair mating by examining the nature of male counter-adaptations to female extra-pair mating. If, historically, women sought extra-pair mating, selection should have forged male counter-adaptations (the shifts in male proprietariness across the female cycle purportedly being examples; see Buss, 2000). The kinds of counter-adaptation that have evolved may provide insight into the threats to men posed by partner infidelity in ancestral environments. Shackelford and colleagues have aggressively pursued this tack (see Chapter 4 in this volume). Impressively, for instance, they found that, as men experience greater risk from partners having engaged in extra-pair sex as a function of time separated from them, they show greater attraction to partners and interest in having sex with them, independent of confounding variables such as time since last sex (Shackelford et al., 2002). Another study (Goetz et al., 2005) found that men purportedly at recurrent risk of extra-pair mating (e.g. owing to partners' personality characteristics) engaged in particular copulatory behaviors speculated to displace semen (e.g. deep, vigorous thrusting) more so than men at less risk of extra-pair mating (see also Gallup et al., 2003). In another ingenious line of research, Platek et al. (2003, 2004) found that men respond differently to self-resemblance of children's faces than do women. In particular, in hypothetical investment decisions, men offer less investment to children who look less like themselves than to children whose faces have been manipulated to look like the men themselves; women do not differentially respond as a function of self-resemblance. Purportedly, men's behavior reflects anti-cuckoldry adaptation. These results are fascinating. Currently, they do not point to any particular benefit of infidelity for females ancestrally. Indeed, they need not have arisen in the context of any adaptation for extra-pair mating on the part of females. Even if ancestral women engaged in extra-pair sex due to a byproduct of other aspects of their mating psychology (or even sexual coercion on the part of men), men could have evolved anti-cuckoldry adaptations in response. Research exploring the precise nature of anti-cuckoldry adaptations in men (e.g. the circumstances in which men are particularly vigilant to children's self-resemblance or respond to time apart from their partner) may reveal more specifically the circumstances in which men historically were threatened by female extra-pair mating, which could in turn inform our understanding about existence of female adaptations for extra-pair mating.

Third, key evidence of historical male–female conflicts of interest may be found in morphological or physiological adaptations that may have evolved in the context of sexually antagonistic coevolution. Gallup et al. (2003), for instance, have proposed that the penis (specifically, the coronal ridge) has been shaped to function as a semen-displacement device in the context of female extra-pair mating. While the speculations and data presented are intriguing, evidence that the penis has been shaped by such selection would be strengthened by evidence that female reproductive tracts have been shaped through sexually antagonistic coevolution to reduce the effectiveness of the penis as a semen-displacement device. Alternatively, penis shape may have been selected through traditional sexual selection by female choice; this possibility should be ruled out. (For pertinent discussions of the issues and kind of evidence needed to make claims in favor of sexually antagonistic evolution, see Eberhard, 2004, and Hosken & Stockley, 2004.)

Seminal fluid has many compounds in it (e.g. a variety of immunomodulators) that presumably evolved because they enhance male reproductive success (see Gangestad, Thornhill, & Garrer-Apgor, 2005a, for a brief review). It is often assumed (at least implicitly) in writings about these compounds that they aid in the female's interest as well by increasing the chances of sperm surviving and reaching the egg in the face of an environment (the female reproductive tract) hostile to foreign bodies. Another possibility, however, is that at least some may have evolved through conflicts between the sexes. A variety of compounds in seminal fluid are not immunosuppressive in general but rather bias immune response away from targeting cell-mediated response (generally against viruses and some bacteria) and toward humoral responses (generally against larger pathogens). A female immune response against sperm (perhaps facilitated by these immunomodulators) actually appears to increase the odds of favorable outcomes (e.g. proper implantation of a conceptus; e.g. Robertson, Bromfield, & Tremellen, 2003). Interestingly, however, the female immune system appears to be biased away from humoral activity at highest conception risk, opposite of the effects of seminal fluid immunomodulators (e.g. Franklin & Kutteh, 1999; Gravitt et al., 2003). The opponent effects could possibly be the result of sexually antagonistic coevolution.

If male fertility is enhanced by female immune recognition of sperm antigens, selection should favor seminal products that enhance that recognition (through humoral immunoactivity), possibly as a cryptic means of favoring MHC-compatible fathers. The effects of those products may not optimize female immune functioning, cryptic choice, investment in non-optimal offspring, and so on. Small differences in genetic interests between the sexes could drive a coevolutionary process through which male seminal products are selected for their powerful effects on female immune responsivity and the female reproductive tract is selected to counter these effects. That is, some males' seminal products (and, possibly, proteins on the sperm acrosome) may more effectively lead to conception and be sexually selected, whereas some females may be better able to control activities in their reproductive tract in their own interests, leading to selection for resistance. In such a chase-away selection process, males who are most fit overall may have an edge in producing the most effective sperm (e.g. Kokko *et al.*, 2003). Hence, sexual selection of this sort may ultimately be an instance of good-genes sexual selection. If so, then female resistance is maintained by the fact that men whose semen is best able to overcome it offer genetic benefits and thereby functions as a cryptic choice mechanism for good genes (see Gangestad *et al.*, 2005a, for a more extensive discussion).

It does not seem plausible that human females would pay the costs of extra-pair copulation simply to run sperm competition races blind to explicit choice based on perceived features. Hence, the possibility that differences in semen properties themselves importantly drive female extra-pair mating behavior seems highly unlikely, though they could augment selection for extra-pair copulation to obtain genetic benefits. The design features of seminal fluid compounds and responses in the female reproductive tract are not yet well understood and, hence, these ideas are highly speculative at this point. (Human reproductive biologists have not entertained ideas such as these to date. Rice's [1996] highly important paper on sexually antagonistic coevolution, in which he discussed antagonistic effects of seminal fluid, has been cited about 300 times to date, but never in a human reproductive biology journal.) The general point is that, ultimately, understanding of these systems could offer clues about the nature of ancestral male–female relations, including female infidelity.

Gallup, Burch, & Platek (2002) present data suggesting that exposure to seminal fluid affects women's mood. In particular, women who use condoms have more depressive symptoms than women who are exposed to semen because they use other means of birth control and the level of depressive symptoms covaries with time since last exposure within the latter group. Assuming these effects hold up after additional attempts to control for potential confounding variables (e.g. personality differences between women who do and do not use alternatives to condoms), a key question is whether these effects have themselves been functional in human evolution or are byproducts of other functions of semen properties (e.g. immunomodulation) and, if the former, what selective pressures led to these effects.

Summary

Do women possess adaptation that functions in the context of extra-pair mating? Have human males evolved counter-adaptations? In this chapter, I have discussed several themes pertinent to these questions. First, there appears to be a nearly inevitable genetic conflict of interest over from whom a female obtains genes for offspring between pairs of individuals who biparentally invest whenever there are neighboring males. Females could potentially gain from extra-pair mating and male counter-responses (including discriminative investment based on likelihood of paternity) are probably the main force suppressing those interests. These counter-responses, however, evolve through sexually antagonistic coevolution. Second, then, the EPP rate need not be high for sexually antagonistic coevolution fueled by conflict over where females obtain genes for offspring to occur. Third, the best evidence for female adaptation for extra-pair mating and male counter-adaptation is to be found in the design features of purported adaptations. Currently available evidence is not sufficient to establish firmly the existence of female adaptations for extra-pair mating. Lines of research on such adaptations and male counter-responses have been fruitfully pursued, however. Even if not providing conclusive evidence for sexually antagonistic coevolution, this research has generated a substantial number of intriguing findings about human sexual relations.

References

Adams, D. B., Gold, A. R., and Burt, B. A. (1978). Rise in female-initiated sexual activity at ovulation and its suppression by oral contraceptives. *New England Journal of Medicine*, **299**, 1145–50.

Alexander, R. D. and Noonan, K. (1979). Concealment of ovulation, parental care, and human social evolution. In N. A. Chagnon and W. Irons, eds., *Evolutionary Biology and Human Behavior: an Anthropological Perspective*. North Scituate, MA: Duxbury, pp. 402–35.

Bancroft, J., Sanders, D., Davidson, D., and Warner, P. (1983). Mood-sexuality, hormones, and the menstrual cycle. III. Sexuality and the role of androgens. *Psychosomatic Medicine*, **45**, 509–16.

Bellis, M. A. and Baker, R. R. (1990). Do females promote sperm competition? Data for humans. *Animal Behaviour*, **40**, 997–9.

Birkhead, T. R. and Møller, A. P. (1995). Extrapair copulation and extra-pair paternity in birds. *Animal Behaviour*, **49**, 843–8.

Buller, D. J. (2005). *Adapting Minds*. Cambridge, MA: MIT Press.

Burt, A. (1995). Perspective: The evolution of fitness. *Evolution*, **49**, 1–8.

Buss, D. M. (2000). *Dangerous Passions*. New York: Free Press.

Charlesworth, B. (1990). Mutation-selection balance and the evolutionary advantage of sex and recombination. *Genetical Research*, **55**, 199–221.

Charlesworth, B. and Hughes, K. A. (1998). The maintenance of genetic variation in life history traits. In R. S. Singh and C. B. Krimbas, eds., *Evolutionary Genetics from Molecules to Morphology*. Cambridge: Cambridge University Press.

Eberhard, W. G. (2004). Rapid divergent evolution of sexual morphology: Comparative tests of antagonistic coevolution and traditional female choice. *Evolution*, **58**, 1947–70.

Franklin, R. D. and Kutteh, W. H. (1999). Characterization of immunoglobulins and cytokines in human cervical mucus: influence of exogenous and endogenous hormones. *Journal of Reproductive Immunology*, **42**, 93–106.

Gallup, G. G., Burch, R. L., and Platek, S. M. (2002). Does semen have antidepressant properties? *Archives of Sexual Behavior*, **31**, 289–93.

Gallup, G. G., Burch, R. L., Zappieri, M. L., Parvez, R. A., Stockwell, M. L., and Davis, J. A. (2003). The human penis as a semen displacement device. *Evolution and Human Behavior*, **24**, 277–89.

Gangestad, S. W. and Thornhill, R. (1998). Menstrual cycle variation in women's preference for the scent of symmetrical men. *Proceedings of the Royal Society of London B*, **262**, 727–33.

Gangestad, S. W., Thornhill, R., and Garver, C. E. (2002). Changes in women's sexual interests and their partners' mate retention tactics across the menstrual cycle: Evidence for shifting conflicts of interest. *Proceedings of the Royal Society of London B*, **269**, 975–82.

Gangestad, S. W., Simpson, J. A., Cousins, A. J., Garver-Apgar, C. E., and Christensen, P. N. (2004). Women's preferences for male behavioral displays change across the menstrual cycle. *Psychological Science*, **15**, 203–7.

Gangestad, S. W., Thornhill, R., and Garver-Apgar, C. E. (2005a). Adaptations to ovulation. In D. M. Buss, ed., *Handbook of Evolutionary Psychology*. New York: Wiley.

Gangestad, S. W., Thornhill, R., and Garver-Apgar, C. E. (2005b). Female sexual interests across the ovulatory cycle depend on primary partner developmental instability. *Proceedings of the Royal Society of London B* (in press).

Goetz, A. T., Shackelford, T. K., Weekes-Shackelford, A., *et al.* (2005). Mate retention, semen displacement, and human sperm competition: a preliminary investigation of tactics to prevent and correct female infidelity. *Personality and Individual Differences*, **38**, 749–63.

Grammer, K. (1993). 5-α-Androst-16en-3α-on: a male pheromone? A brief report. *Ethology and Sociobiology*, **14**, 201–14.

Gravitt, P. E., Hildesheim, A., Herrero, R., *et al.* (2003). Correlates of IL-10 and IL-12 concentrations in cervical secretions. *Journal of Clinical Immunology*, **23**, 175–83.

Greiling, H. and Buss, D. M. (2000). Women's sexual strategies: the hidden dimension of short-term extra-pair mating. *Personality and Individual Differences*, **28**, 929–63.

Haig, D. (1993). Genetic conflicts in human pregnancy. *Quarterly Review of Biology*, **68**, 495–532.

Hamilton, W. D. and Zuk, M. (1982). Heritable true fitness and bright birds: a role for parasites. *Science*, **218**, 384–7.

Haselton, M. G. and Miller, G. F. (2005). Evidence for ovulatory shifts in attraction to artistic and entrepreneurial excellence. *Human Nature* (in press).

Hedrick, P. W. and Black, F. L. (1997). HLA and mate selection: no evidence in South Amerindians. *American Journal of Human Genetics*, **61**, 505–11.

Hendricks, C., Piccinino, L. J., Udry, J. R., Chimbira, T. H. K. (1987). Peak coital rate coincides with onset of lutenizing hormone surge. *Fertility and Sterility*, **48**, 234–238.

Holland, B. and Rice, W. R. (1999). Experimental removal of sexual selection reverses intersexual antagonistic coevolution and removes a reproductive load. *Proceedings of the National Academy of Sciences USA*, **96**, 5083–8.

Hosken, D. J. and Stockley, P. (2004). Sexual selection and genital evolution. *Trends in Ecology and Evolution*, **19**, 87–93.

Houle, D. (1992). Comparing evolvability and variability of traits. *Genetics*, **130**, 195–204.

Hrdy, S. B. (1979). Infanticide among animals: a review, classification, and examination of the implications for the reproductive strategies of females. *Ethology and Sociobiology*, **1**, 13–40.

Hummel, T., Gollisch, R., Wildt, G., and Kobal, G. (1991). Changes in olfactory perception during the menstrual cycle. *Experientia*, **47**, 712–15.

Ihara, Y., Aoki, K., Tokumaga, K., Takahashi, K., and Juji, T. (2000). HLA and human mate choice: tests on Japanese couples. *Anthropological Science*, **108**, 199–214.

Jennions, M. D. and Petrie, M. (2000). Why do females mate multiply?: A review of the genetic benefits. *Biological Reviews*, **75**, 21–64.

Johnston, V. S., Hagel, R., Franklin, M., Fink, B., and Grammer, K. (2001). Male facial attractiveness: evidence for hormone mediated adaptive design. *Evolution and Human Behavior*, **23**, 251–67.

Kappeler, P. M. and van Schaik, C. P. (eds.). (2003). *Sexual Selection in Primates: New and Comparative Perspectives*. Cambridge: Cambridge University Press.

Kokko, H., Brooks, R., Jennions, M. D., and Morley, J. (2003). The evolution of mate choice and mating biases. *Proceedings of the Royal Society of London B*, **270**, 653–64.

Matteo, S. and Rissman, E. F. (1984). Increased sexual activity during the midcycle portion of the human menstrual cycle. *Hormones and Behavior*, **18**, 249–55.

Penn, D. J. and Potts, W. K. (1999). The evolution of mating preferences and major histocompatibility complex genes. *American Naturalist*, **153**, 145–64.

Penton-Voak, I. S., Perrett, D. I., Castles, D., Burt, M., Koyabashi, T., and Murray, L. K. (1999). Female preference for male faces changes cyclically. *Nature*, **399**, 741–2.

Persky, H., Charney, N., Leif, H. I., O'Brien, C. P., Miller, W. R., and Strauss, D. (1978). The relationship of plasma estradiol to sexual behavior in young women. *Psychosomatic Medicine*, **40**, 523–37.

Petrie, M. and Kempenaers, B. (1998). Extra-pair paternity in birds: Explaining variation between species and populations. *Trends in Ecology and Evolution*, **13**, 52–8.

Pillsworth, E. G., Haselton, M. G., and Buss, D. M. (2004). Ovulatory shifts in female sexual desire. *Journal of Sex Research* (in press).

Platek, S. M., Critton, S. R., Burch, R. L., Frederick, D. A., Myers, T. E., and Gallup, G. G. (2003). How much paternal resemblance is enough? Sex differences in hypothetical investment decisions but not in the detection of resemblance. *Evolution and Human Behavior*, **24**, 81–7.

Platek, S. M., Raines, D. M., Gallup, G. G. *et al.* (2004). Reactions to children's faces: Males are more affected by resemblance than females are, and so are their brains. *Evolution and Human Behavior*, **25**, 394–405.

Pomiankowski, A. and Møller, A. P. (1995). A resolution of the lek paradox. *Proceedings of the Royal Society of London B*, **260**, 21–9.

Putz, D. (2005). Menstrual phase and mating context affect women's preferences for male voice pitch. *Evolution and Human Behavior* (in press).

Regan, P. C. (1996). Rhythms of desire: the association between menstrual cycle phases and female sexual desire. *Canadian Journal of Human Sexuality*, **5**, 145–56.

Rice, W. R. (1988). Heritable variation as a prerequisite for adaptive female choice: the effect of mutation-selection balance. *Evolution*, **42**, 817–20.

Rice, W. R. (1996). Sexually antagonistic male adaptation triggered by experimental arrest of female evolution. *Nature*, **381**, 232–4.

Rice, W. R. and Holland, B. (1997). The enemies within: intragenomic conflict, interlocus contest evolution (ICE), and the intraspecific Red Queen. *Behavioral Ecology and Sociobiology*, **41**, 1–10.

Rikowski, A. and Grammer, K. (1999). Human body odour, symmetry and attractiveness *Proceedings of the Royal Society of London B*, **266**, 869–74.

Robertson, S. A., Bromfield, J. J., and Tremellen, K. P. (2003). Seminal 'priming' for protection from preeclampsia: a unifying hypothesis. *Journal of Reproductive Immunology*, **59**, 253–65.

Schreinerss-Engel, P., Schiavi, R. C., Smith, H., and White, D. (1981). Sexual arousability and the menstrual cycle. *Psychosomatic Medicine*, **43**(3), 199–214.

Shackelford, T. K., LeBlanc, G. J., Weekes-Shackelford, V. A., Bleske-Rechek, A. L., Euler, H. A., and Hoier, S. (2002). Psychological adaptation to human sperm competition. *Evolution and Human Behavior*, **12**, 123–38.

Thornhill, R. and Gangestad, S. W. (1999). The scent of symmetry: a human pheromone that signals fitness? *Evolution and Human Behavior*, **20**, 175–201.

Thornhill, R., Gangestad, S. W., Miller, R., Scheyd, G., McCollough, J., and Franklin, M. (2003). MHC, symmetry and body scent attractiveness in men and women (*Homo sapiens*). *Behavioral Ecology* (in press).

Trivers, R. L. (1972). Parental investment and sexual selection. In B. Campbell, ed., *Sexual Selection and the Descent of Man, 1871–1971*. Chicago, IL: Aldine.

Wedekind, C. and Füri, S. (1997). Body odor preference in men and women: do they aim for specific MHC combinations or simply heterozygosity? *Proceedings of the Royal Society of London B*, **264**, 1471–79.

Wedekind, C., Seebeck, T., Bettens, F., and Paepke, A. J. (1995) MHC-dependent mate preferences in humans. *Proceedings of the Royal Society of London B*, **260**, 245–9.

Wilcox, A. J., Baird, D. D., Dunson, D. B., McConnaughey, D. R., Kesner, J. S., and Weinberg, C. R. (2004). On the frequency of intercourse around ovulation: evidence for biological influences. *Human Reproduction*, **19**, 1539–43.

Williams, G. C. (1992). *Natural Selection: Domains, Levels, and Challenges*. New York: Oxford University Press.

Zeh, J. A. and Zeh, D. W. (2001) Spontaneous abortion depresses female sexual receptivity in a viviparous arthropod. *Animal Behaviour*, **62**, 427–33.

4

Predicting violence against women from men's mate-retention behaviors

TODD K. SHACKELFORD AND AARON T. GOETZ
Florida Atlantic University

Introduction

Male sexual jealousy is a frequently cited cause of non-lethal and lethal violence in romantic relationships (e.g. Buss, 2000; Daly & Wilson, 1988; Daly, Wilson, & Weghorst; 1982; Dutton, 1998). Evolutionary psychologists hypothesized two decades ago that male sexual jealousy may have evolved to solve the adaptive problem of paternity uncertainty (Daly *et al.*, 1982; Symons, 1979). Unlike women, men face uncertainty about the paternity of their children because fertilization occurs within women. Without direct cues to paternity, men risk cuckoldry, and therefore might unwittingly invest in genetically unrelated offspring. Cuckoldry is a reproductive cost inflicted on a man by a woman's sexual infidelity or temporary defection from her regular long-term relationship. Ancestral men also would have incurred reproductive costs by a long-term partner's permanent defection from the relationship. These costs include loss of the time, effort, and resources the man has spent attracting his partner, the potential misdirection of his resources to a rival's offspring, and the loss of his mate's investment in offspring he may have had with her in the future (Buss, 2000).

Expressions of male sexual jealousy historically may have been functional in deterring rivals from mate poaching (Schmitt & Buss, 2001) and deterring a mate from a sexual infidelity or outright departure from the relationship (Buss *et al.*, 1992; Daly *et al.*, 1982; Symons, 1979). Buss (1988) categorized the behavioral output of jealousy into different "mate-retention" tactics, ranging from vigilance over a partner's whereabouts to violence against rivals (see also Buss & Shackelford, 1997). Performance of these tactics is assessed by the Mate

Retention Inventory (MRI; Buss, 1988). Buss's taxonomy (1988) partitioned the tactics into two general categories: *intersexual manipulations* and *intrasexual manipulations*. Intersexual manipulations include behaviors directed toward one's partner, and intrasexual manipulations include behaviors directed toward same-sex rivals. Intersexual manipulations include direct guarding, negative inducements, and positive inducements. Intrasexual manipulations include public signals of possession.

Because male sexual jealousy has been linked to violence in relationships, and because mate-retention tactics are behavioral manifestations of jealousy, men's use of these tactics is predicted to be associated with violence toward their partners. Indeed, Buss and Shackelford (1997) hypothesized that the use of some mate-retention tactics may be early indicators of violence in romantic relationships. Unfortunately, little is known about *which specific acts and tactics* of men's mate-retention efforts are linked with violence. The primary exception is the study by Wilson, Johnson, and Daly (1995), which identified several predictors of partner violence – notably, verbal derogation of the mate and attempts at sequestration such as limiting access to family, friends, and income. This chapter highlights some of our recent research (see Shackelford *et al.*, 2005), which was designed to identify specific behaviors that portend violence in romantic relationships, and to contribute to a better understanding of violence against women. Identifying the predictors of partner violence would be theoretically valuable, and may provide information relevant to developing interventions designed to reduce partner violence or to help women avoid such violence.

Assessing violence in romantic relationships

Dobash *et al.* (1995, 1996) developed three indexes to assess the occurrence and consequences of violence in relationships. Violence toward partners is not limited to physical assaults, but also includes nonphysical controlling and coercive behaviors. To measure the occurrence of nonphysical controlling and coercive behaviors in relationships, Dobash *et al.* (1996) developed the Controlling Behavior Index (CBI), which includes assessments of threats, psychological maltreatment, and verbal violence. The Violence Assessment Index (VAI; Dobash *et al.*, 1995) measures specific methods of assault, objects used in assaults, and parts of the body to which assaults are directed. The types of violence assessed range from pushing to choking. Because the effects of violence can range from minor wounds (e.g. a scratch) to more severe damage (e.g. an internal injury), Dobash *et al.* (1995) developed the Injury Assessment Index (IAI) to measure the physical consequences of violence against partners. The IAI is comprehensive in that it measures the specific injury (e.g. bruise, cut) and the location of the injury on the body (e.g. face, limb).

Predictors of violence in romantic relationships

DIRECT GUARDING

Tactics within the direct guarding category of the MRI include vigilance, concealment of mate, and monopolization of time. An exemplary act for each tactic is, respectively, "He dropped by unexpectedly to see what she was doing," "He refused to introduce her to his same-sex friends," and "He monopolized her time at the social gathering." Each of these tactics implicates what Wilson and Daly (1992) term "male sexual proprietariness," which refers to the sense of entitlement men sometimes feel that they have over their partners and, more specifically, their partners' sexual behavior. Male sexual proprietariness motivates behaviors in men designed to regulate and restrict women's sexual autonomy. A sexually proprietary male psychology has been proposed to be an adaptive solution to the problems of intrasexual competition for mates and cuckoldry (Buss *et al.*, 1992; Daly *et al.*, 1982; Symons, 1979). Ancestral men who attempted to limit their partners' sexual autonomy were likely to have been more reproductively successful because, on average, they were better able to deter rivals from encroaching and to deter mates from straying, than were men who made no such attempts. From a woman's point of view, however, these mate-guarding actions may inflict costs on her by restricting her freedom of sexual choice, restricting her mobility, limiting her social contacts, and impeding her ability to pursue her own interests unfettered.

Wilson *et al.* (1995) demonstrated that violence against women is linked closely to their partners' autonomy-limiting behaviors. Women who affirmed items such as "He is jealous and doesn't want you to talk to other men," were more than twice as likely to have experienced serious violence by their partners. Of those women who were questioned further about their experiences with serious violence, 56% reported being fearful for their lives and 72% required medical attention following an assault. Because direct guarding is associated specifically with men's autonomy-limiting behaviors, we expected direct guarding to be related positively to violence in romantic relationships.

INTERSEXUAL NEGATIVE INDUCEMENTS

In addition to direct guarding, men attempt to retain their partners by using intersexual negative inducements. Punish mate's infidelity threat, for example, includes acts such as "He yelled at her after she showed interest in another man." This tactic has a violent theme and, therefore, we expected it to be related positively to violence in relationships. Because jealousy is a primary cause of violence against women, those women who openly threaten infidelity, consequently inducing jealousy in their partners, are predicted to be more likely to suffer violence at the hands of their partners.

POSITIVE INDUCEMENTS

Not all mate-retention tactics are expected to predict positively violence toward partners. Some of these tactics include behaviors that are not in conflict with a romantic partner's interests and, indeed, may be encouraged and welcomed by a partner (Buss, 1988, 2000). One might not expect, for example, that men who attempt to retain their partners by using positive inducements will behave more violently toward their partners than men who do not deploy such tactics. For example, men who affirm love and care acts (e.g. "I was helpful when she really needed it") and resource display acts (e.g. "I bought her an expensive gift") may not be expected to use violence against their partners. Men who have resources might be able to retain their partners using methods that are not available to men lacking resources. Indeed, Daly and Wilson (1988) predicted that men who cannot retain mates through positive inducements may be more likely to resort to violence. Following Daly and Wilson (1988), we expected the use of positive inducements to be related negatively to female-directed violence.

PUBLIC SIGNALS OF POSSESSION

Tactics within the public signals of possession category include verbal possession signals (e.g. "He mentioned to other males that she was taken"), physical possession signals (e.g. "He held her hand when other guys were around"), and possessive ornamentation (e.g. "He hung up a picture of her so others would know she was taken"). Public signals of possession reflect male sexual proprietariness and, therefore, we expected the use of public signals of possession to be related positively to female-directed violence.

Shackelford *et al.* (2005) collected data using Buss's (1988) MRI to measure female-directed mate-retention behaviors, and Dobash *et al.*'s (1995, 1996) CBI, VAI, and IAI to measure female-directed controlling behaviors, violence, and injuries, respectively. We generated four predictions derived from the hypothesis that men's use of mate-retention tactics is variably associated with violence against their partners.

> *Prediction 1*: men's use of direct guarding will be related *positively* to their use of controlling behaviors (Prediction 1.1), violence (1.2), and injuries inflicted on their partners (1.3).

Prediction 2: men's use of intersexual negative inducements will be related *positively* to their use of controlling behaviors (2.1), violence (2.2), and injuries inflicted on their partners (2.3).

Prediction 3: men's use of positive inducements will be related *negatively* to their use of controlling behaviors (3.1), violence (3.2), and injuries inflicted on their partners (3.3).

Prediction 4: men's use of public signals of possession will be related *positively* to their use of controlling behaviors (4.1), violence (4.2), and injuries inflicted on their partners (4.3).

In study 1 (Appendix 4.1), we collected self-reports from several hundred men about their use of mate-retention tactics and their partner-directed violence in a current romantic relationship. Men and women sometimes are discordant about instances of violence in their relationship (e.g. Dobash *et al.*, 1998; Magdol *et al.*, 1997). The consensus among researchers is that men underreport the violence they inflict on their partners, whereas women report this violence with relative accuracy. Because women's reports of violence in relationships reflect more accurately the incidence of such violence, study 2 (Appendix 4.1) secured women's reports of their partners' use of mate-retention tactics and partner-directed violence. For reportorial efficiency, we report the conduct and results of studies 1 and 2 together. We then report the results of a third study (Appendix 4.2) in which the linked responses of husbands and their wives were used to conduct additional tests of the four predictions.

General discussion

Some mate-retention tactics often are welcomed by their recipients. Holding his partner's hand in public, for example, may signal to a woman her partner's commitment and devotion to her. Frequent use of some tactics of commitment and devotion, however, may also be harbingers of violence against a romantic partner. The current studies examined how mate-retention tactics are related to violence in romantic relationships, using the reports of independent samples of several hundred men and women in committed, romantic relationships (studies 1 and 2; Appendix 4.1), and using the reports of 107 married men and women (study 3; Appendix 4.2).

We hypothesized that, because male sexual jealousy is a primary cause of violence in romantic relationships, and because mate-retention tactics are behavioral manifestations of jealousy, men's use of mate-retention tactics will be associated with female-directed controlling behaviors, violence, and injuries. We derived and tested four predictions from this hypothesis: men's use of direct guarding, intersexual negative inducements, and public signals of possession will be related *positively* to female-directed control, violence, and injuries (predictions 1, 2, and 4, respectively); men's use of positive inducements, in contrast, will be related *negatively* to female-directed control, violence, and injuries (prediction 3).

Predictions 1 and 2 were supported by the data collected in study 1. According to men's self-reports, their use of direct guarding and intersexual negative inducements is related positively to controlling behaviors, violence, and injuries (predictions 1 and 2, respectively). One facet of prediction 4 was supported by the data in study 1: men's self-reported use of public signals of possessions is related positively to their controlling behaviors. In addition, men who reported using frequently the tactics of emotional manipulation, punish mate's infidelity threat, monopolization of time, derogation of competitors, jealousy induction, and vigilance also reported inflicting more violence on their partners.

Predictions 1 and 2 also were supported by the data collected in study 2. According to women's reports of their partners' behaviors, men's use of direct guarding and intersexual negative inducements was related positively to female-directed controlling behaviors, violence, and injuries (predictions 1 and 2, respectively). Paralleling the results of Study 1, one facet of prediction 4 was supported by the data in study 2: men's use of public signals of possession was related positively to their controlling behaviors. In addition, women who reported that their partners use frequently the tactics concealment of mate, emotional manipulation, vigilance, monopolization of time, and punish mate's infidelity threat also reported more violence in their relationships.

Predictions 1, 2, and 4 were supported by the data collected in study 3. According to husbands' reports of their mate-retention tactics and their wives' reports of violence, husbands' use of direct guarding, intersexual negative inducements, and public signals of possession were related positively to female-directed violence (predictions 1, 2, and 4, respectively). In addition, husbands who reported using frequently the tactics vigilance, emotional manipulation, monopolization of time, possessive ornamentation, and concealment of mate had wives' who report more violence in their relationships.

With few exceptions, we found the same pattern of results using three independent samples. Moreover, these samples were not just independent, but provided different perspectives (the male perpetrator's, the female victim's, and a combination of the two) on the same behaviors – men's mate-retention behaviors and men's violence against their partners. We identified overlap between the best predictors of violence across the studies. For example, men's use of emotional manipulation, monopolization of time, and punish mate's infidelity threat are among the best predictors of female-directed violence, according to independent reports provided by men and women, and according to reports provided by husbands and their wives. The three perspectives also converged on which tactics are the weakest predictors of relationship violence. For example, love and care and resource display are among the weakest predictors of female-directed violence. These parallel patterns of results provide corroborative support for the hypothesis that men's use of certain mate-retention tactics is associated with female-directed violence.

Some mate-retention behaviors involve the provisioning of benefits rather than the infliction of costs (Buss, 1988; Buss & Shackelford, 1997). Prediction 3 was designed to test Daly and Wilson's (1988) hypothesis that men who are unable to employ positive inducements such as gift-giving and the provisioning of material resources to retain a mate will be more likely to use violence as a means of mate retention. Violence against their partners therefore was predicted to be related negatively to men's use of positive inducements. The current research provides no support for this prediction and, in fact, provides some evidence for the reverse relationship. Across the three studies, the significant correlations identified between positive inducements and controlling behavior, violence, and injuries are exclusively positive. A *post hoc* speculation for these results is that men faced most severely with the adaptive problem of a partner's defection may ratchet up their use of all mate-retention tactics, both positive (benefit provision) and negative (cost infliction).

MATE-RETENTION TACTICS AS PREDICTORS OF RELATIONSHIP VIOLENCE

The tactic of emotional manipulation was the highest-ranking predictor of violence in romantic relationships in study 1, and the second highest-ranking predictor in studies 2 and 3. The items that comprise the emotional manipulation tactic include, "He told her he would 'die' if she ever left," and "He pleaded that he could not live without her." Such acts seem far removed from those that might presage violence. The robust relationship between female-directed violence and men's use of emotional manipulation can be interpreted in at least two ways. Emotional manipulation may be a post-violence "apologetic" tactic. Perhaps men who behave violently toward their partners are apologizing and expressing regret for their violent behavior. Indeed, Walker (2000) has observed that, following a violent episode, men often are apologetic, expressing remorse and pleading for forgiveness.

Another possibility is that emotional manipulation may occur before relationship violence, making it a true harbinger of violence. Perhaps a man who tells his partner that he would die if she ever left him has invested so heavily in the relationship and perceives that he has so much to lose if the relationship ended that he reacts violently when the relationship is threatened. Men who are of much lower mate value than their partners, for example, may have so much to lose that they become violent when their partner defects temporarily (i.e. commits a sexual infidelity) or permanently (i.e. ends the relationship). Future research would benefit from determining whether the use of emotional manipulation occurs before or after relationship violence. A longitudinal study, for example, could assess men's use of mate-retention tactics in the beginning of a relationship and then subsequently assess men's violence against their partners. If men who became violent toward their partners as the relationship progressed did not use emotional manipulation at the start of the relationship but only after they became violent, this would suggest that emotional manipulation may be an apologetic tactic used to seek forgiveness for a violent transgression.

Monopolization of time also ranked as a strong predictor of violence across the three studies. Example acts included in this tactic are "He spent all his free time with her so that she could not meet anyone else" and "He would not let her go out without him." The positive relationships identified in the current studies between monopolization of time and violence is consistent with the demonstration by Wilson *et al.* (1995) that violence against women is linked closely to their partners' autonomy-limiting behaviors. Wilson *et al.* (1995) found that women who affirmed items such as "He tries to limit your contact with family or friends" are twice as likely to have experienced serious violence by their partners.

We identified significant correlations between the mate-retention tactic sexual inducements and relationship violence in studies 2 and 3. Sexual inducements includes items such as "He gave in to her sexual requests," and "He performed sexual favors to keep her around." Guided by sperm-competition theory (Parker, 1970), Goetz *et al.* (2005) found that men partnered to women who are more likely to be sexually unfaithful are also more likely to perform sexual inducements to retain their partners. Goetz *et al.* (2005) interpreted a man's use of sexual inducements to be a "corrective" tactic designed to place his sperm in competition with rival sperm that may be present in his partner's reproductive tract. Men's use of sexual inducements and female-directed violence are both motivated by sexual jealousy (Daly & Wilson, 1988; Daly *et al.*, 1982; Goetz *et al.*, 2005), and this may account for the consistent relationships between men's use of sexual inducements and female-directed violence.

MATE-RETENTION ACTS AS PREDICTORS OF RELATIONSHIP VIOLENCE

The highest-ranking correlations between single acts and relationship violence are not particularly consistent across the three studies. The data of studies 1 and 2 were secured from a single data source (men and women, respectively). The data of study 3 arguably have greater credibility, because reports of mate retention and violence were provided by different data sources. For this reason, and for reportorial efficiency, we limit our discussion of the results of act-level analyses to study 3. More specifically, we discuss three of the highest-ranking correlations between single acts of mate retention and violence based on husbands' reports of their mate retention and their wives' reports of violence.

The acts "Dropped by unexpectedly to see what my partner was doing" and "Called to make sure my partner was where she said she would be" are the third and fifth highest-ranking predictors of violence, respectively. These acts are included in the tactic of vigilance, which is the highest-ranking tactic-level predictor of violence in study 3. Given that (1) two of the top five act-level predictors of violence are acts of vigilance, (2) the numerically best tactic-level predictor of violence is vigilance, and (3) seven of the nine acts included within the vigilance tactic are correlated significantly with violence (correlations available upon request), a man's vigilance over his partner's whereabouts is likely to be a key signal of his partner-directed violence. The acts within the vigilance tactic are examples of autonomy-limiting behaviors – behaviors motivated by male sexual proprietariness and designed to restrict women's sexual autonomy (Wilson & Daly, 1992). Wilson *et al.* (1995) demonstrated that men's use of autonomy-limiting behaviors is associated with female-directed violence. Wilson *et al.* (1995) found that 40% of women who affirmed the statement "He insists on knowing who you are with and where you are at all times" reported experiencing serious violence at the hands of their husbands. The vigilance acts highlighted above contain both the *who* and the *where* components of the Wilson *et al.*'s (1995) statement regarding a partner's autonomy-limiting behaviors.

The act "Told my partner that I would 'die' if my partner ever left" is the fourth highest-ranking predictor of violence. This act is included in the tactic of emotional manipulation, which is the second highest-ranking tactic-level predictor of violence in study 3. It is not known whether a man who affirms this item is attempting to persuade his wife not to end the relationship because he committed some abhorrent act, such as partner violence, or might be telling his wife this because he is of much lower mate value than she and, therefore, would have much to lose if the relationship ended. In the former interpretation the act is a consequence of violence and in the latter violence is a consequence of a threat to the valued relationship. Future research should examine whether this and other acts of emotional manipulation occur before or after violence has occurred.

Concluding remarks

Mates gained must be retained to actualize the promise inherent in the initial mate selection and successful courting. Mate poaching, infidelity, and defection from a mateship undoubtedly were recurrent adaptive problems over human evolutionary history. Men's psychology of jealousy and the attendant tactics of mate retention appear to be evolved solutions to these adaptive problems. Adaptive solutions need not succeed invariantly; they evolve if they succeed, on average, across the sample space of relevant instances, better than competing designs present in the population at the time. Increased effort devoted to mate retention is predicted to occur when the adaptive problems it was designed to solve are most likely to be encountered – when a mate is particularly desirable, when there exist mate poachers, when there is a mate-value discrepancy, and when the partner displays cues to infidelity or defection (Buss & Shackelford, 1997; Shackelford & Buss, 1997).

Violence directed toward a mate appears to be one manifestation of men's attempts to control a partner and her sexuality. The current studies contribute to knowledge about this pervasive problem on two levels, conceptually and practically. Conceptually, we have identified several expected predictors of men's use of violence that contribute in some measure to a broader theory of men's use of violence. At a practical level, results of these studies can potentially be used to inform women and men, friends and relatives, of danger signs – the specific acts and tactics of mate retention that portend the possibility of future violence in relationships in order to prevent it before it has been enacted.

Appendix 4.1   Studies 1 and 2: men's and women's reports of female-directed mate retention and violence

In three studies, we secured men's and women's reports of men's mate-retention tactics and use of violence in their current romantic relationships. Studies 1 and 2 secured, in independent samples, men's self-reports and women's partner-reports, respectively.

METHODS

Participants

Four hundred and sixty one men and 560 women in a committed, sexual, heterosexual relationship participated in studies 1 and 2, respectively. Participants were drawn from universities and surrounding communities. The mean age of the men was 24.2 years (SD = 7.9 years), the mean age of their partners was 23.2 years (SD = 7.3 years), and the mean length of their relationships was 37.3 months (SD = 59.8 years). The mean age of the women was 21.5 years (SD = 5.4 years), the mean age of their partners was 23.7 years (SD = 6.6 years), and the mean length of their relationships was 28.8 months (SD = 38.05 years). None of the women in study 2 were partners of the men who participated in study 1, making the two studies independent. About half the participants received nominal extra credit toward one of several social science courses in exchange for their participation. The remaining half of participants received credit toward a required research participation component of an intro-ductory psychology course. We did not code for method of data collection, so were unable to include this as a variable in the statistical analyses.

Materials

Participants in both studies completed a survey that included four indexes. The MRI (Buss, 1988) assesses how often men performed 104 mate-retention acts in the past month, ranging from 0 (never) to 3 (often). Previous research has established the reliability, validity, and utility of the MRI as an assessment of mate-retention behaviors (e.g. Buss, 1988; Buss & Shackelford, 1997; Gangestad, Thornhill, & Garver, 2002). The MRI was generated using an act-nomination procedure (e.g. Buss & Craik, 1983) and subsequently refined by the heuristic application of an evolutionary perspective (Buss, 1988). Recent evidence indicates that the tactics identified by Buss (1988) are captured gen-erally by a formal factor analysis (see Gangestad et al., 2002). Even if this were not the case, however, we argue for the continued use of Buss's (1988) mate-retention tactics and super-ordinate categories, which provide continuity with previous work (e.g. Buss, 1988; Buss & Shackelford, 1997; Gangestad et al., 2002; Goetz et al., 2005; Shackelford & Buss, 2000) and, in the context of the current research, organize mate-retention behaviors in a theoretically sensible way that allows for clear tests of the predictions.

The CBI assesses how often men performed 21 controlling acts against their partners in the past month, the VAI how often they performed 26 violent acts against their partners, and the IAI how often their partners sustained each of 20 injuries as a result of their violence against their partners. For each index, responses are recorded using a six-point Likert-type scale anchored by 0 (never) and 5 (11 or more times; Dobash et al., 1995; 1996). Studies by Dobash and colleagues (e.g. 1995, 1996, 1998) have demonstrated the reliability, valid-ity, and utility of the three indexes.

Procedure

To qualify for participation, prospective participants had to be at least 18 years old and currently involved in a committed, sexual, heterosexual relationship. Upon the prospective participant's arrival at the scheduled time and location, the researcher confirmed that the prospective participant met the two participation criteria. If the criteria were met, the researcher handed the participant a consent form, the survey, and two brown security envelopes. The participant was instructed to read and sign the consent form, complete the survey, place the completed survey in one envelope, the consent form in the other envelope, and then seal the envelopes. The participant was instructed to place the sealed envelopes in two boxes – one for surveys, one for consent forms.

RESULTS AND DISCUSSION: MEN'S SELF-REPORTS (STUDY 1)

This section reports the results of seven tests each of the four predictions across three studies (three tests in study 1, three in study 2, and one in study 3 [Appendix 4.2]). We instituted a Bonferroni correction for α inflation that produced a per-prediction corrected and directional α level of $(0.05/7)2 = 0.014$ (see Cohen & Cohen, 1983; Hays, 1988).

To test the predictions, we first standardized with unit-weighting responses to the mate-retention tactics and then averaged the relevant tactics to create the super-ordinate categories defined by Buss (1988). Alpha reliabilities for the tactics and categories were acceptable, ranging from $\alpha = 0.50$ to 0.84, with a mean of 0.71 (see Table 4.1). We then correlated men's scores on the mate-retention categories with their scores on the CBI, VAI, and IAI. For analyses involving tactics and categories, we excluded responses to the mate-retention act "I hit my partner when I caught my partner flirting with someone else" to prevent detection of spurious relationships between mate retention and violence (this exclusion was implemented for parallel analyses in studies 2 and 3).

Consistent with predictions 1.1 and 1.3, men's use of direct guarding correlated positively with their scores on the CBI and IAI; r (413) = 0.41 and 0.14, respectively (both P values < 0.014). Prediction 1.2 was not supported statistically: men's use of direct guarding was positively but not significantly correlated with their scores on the VAI; r (413) = 0.12 (not significant). Consistent with predictions 2.1, 2.2, and 2.3, men's use of intersexual negative inducements correlated positively with their scores on the CBI, VAI, and IAI; r (413) = 0.46, 0.20, and 0.15, respectively (all P values < 0.014). The results did not support prediction 3. Men's use of positive inducements did not correlate negatively with their scores on the CBI, VAI, or IAI; r (413) = 0.22 (P < 0.014), 0.09 (not significant), and 0.05 (not significant), respectively. Consistent with prediction 4.1, men's use of public signals of possession correlated positively with their scores on the CBI; r (413) = 0.20 (P < 0.014). Predictions 4.2 and 4.3 were not supported: men's use of public signals of possession did not correlate positively

Table 4.1. *Study 1: correlations between men's self-reported mate retention and scores on the CBI, VAI, IAI, and OVI.*

Mate-retention category (α)/ mate-retention tactic (α)	CBI	VAI	IAI	OVI	Rank
Direct guarding (0.83)	0.41*	0.12	0.14*	0.16*	
Vigilance (0.82)	0.43*	0.13	0.08	0.12*	7
Concealment of mate (0.67)	0.28*	0.06	0.11	0.10	8
Monopolization of time (0.72)	0.36*	0.14*	0.16*	0.18*	3
Intersexual negative inducements (0.84)	0.46*	0.20*	0.15*	0.20*	
Jealousy induction (0.70)	0.31*	0.11	0.15*	0.16*	5
Punish mate's infidelity threat (0.81)	0.49*	0.19*	0.13	0.19*	2
Emotional manipulation (0.80)	0.42*	0.23*	0.17*	0.24*	1
Commitment manipulation (0.50)	0.20*	0.08	−0.03	0.03	12
Derogation of competitors (0.76)	0.35*	0.13	0.15*	0.17*	4
Positive inducements (0.81)	0.22*	0.09	0.05	0.08	
Resource display (0.84)	0.09	0.06	−0.03	0.02	13
Sexual inducements (0.64)	0.24*	0.03	0.05	0.04	10.5
Appearance enhancement (0.77)	0.16*	0.08	0.03	0.06	9
Love and care (0.66)	0.09	0.04	0.02	0.04	10.5
Submission and debasement (0.68)	0.24*	0.13*	0.12	0.15*	6
Public signals of possession (0.74)	0.20*	0.04	−0.03	0.00	
Verbal possession signals (0.61)	0.17*	0.03	−0.04	−0.01	16
Physical possession signals (0.72)	0.13*	0.02	−0.01	0.01	14
Possessive ornamentation (0.65)	0.20*	0.04	−0.03	0.00	15

Note: $N = 413$; α = alpha reliability. Rank is the rank order of the magnitude of the correlation between the mate-retention tactic and scores on the OVI.
*$P < 0.014$

with their scores on the VAI or IAI; r (413) = 0.04 and −0.03 (both not significant), respectively.

We wanted to identify which tactics and acts best predicted (numerically) the occurrence and consequences of violence in mateships. To simplify the analyses and to obtain a broadband assessment of the relationships of mate retention with violence and injury, we separately standardized with unit-weighting scores on the VAI and IAI and then averaged these standardized scores into a composite Overall Violence Index (OVI; $\alpha = 0.90$). The CBI was not included in the composite variable because several of the constituent items do not directly assess violence. For these exploratory analyses (and parallel analyses in studies 2 and 3, as well as tests of differences between men and women along the target variables), we instituted a liberal adjustment for type I error by reducing

α from 0.05 to 0.01 and implementing two-tailed significance tests (Cohen & Cohen, 1983; Hays, 1988).

We first correlated scores on the mate-retention tactics with scores on the OVI. These correlations are shown in the far-right column of Table 4.1. Emotional manipulation showed the highest-ranking correlation with scores on the OVI, followed by punish mate's infidelity threat, monopolization of time, derogation of competitors, and jealousy induction. Verbal possession signals showed the lowest-ranking correlation with scores on the OVI, followed by possessive ornamentation and physical possession signals. The relationships between men's mate retention and their scores on the VAI and IAI paralleled those with the OVI and, therefore, are not discussed further but are displayed in the middle two columns of Table 4.1. For reportorial completeness, Table 4.1 presents the correlations between men's mate retention and scores on the VAI, IAI, CBI, and OVI.

To identify the specific mate-retention acts that were the best predictors of violence, we computed correlations between each of the mate-retention acts and scores on the OVI. These act-level analyses revealed that 27 of the 104 mate-retention acts correlated significantly and positively with scores on the OVI (these correlations are available upon request). The acts "Cried in order to keep my partner with me," "Told my partner that I would change in order to please her," "Told others my partner was a pain," "Told my partner that the other person they were interested in has slept with everyone," and "Would not let my partner go out without me" were the five highest-ranking correlations (r values $= 0.23, 0.21, 0.21, 0.20,$ and 0.20, respectively; all P values < 0.01). The tactics that include these acts are among the top eight tactic-level predictors of relationship violence.

According to men's self-reports, their use of intersexual negative inducements and direct guarding is related positively to violence against their partners. Also, men who reported using the mate-retention tactics of emotional manipulation, punish mate's infidelity threat, monopolization of time, derogation of competitors, jealousy induction, and vigilance reported the most violence in their relationships. The same pattern of findings emerged when we controlled for the man's age, his partner's age, and the length of their relationship (analyses are available upon request).

RESULTS AND DISCUSSION: WOMEN'S PARTNER-REPORTS (STUDY 2)

As in study 1, we first standardized with unit weighting the mate-retention tactics and then averaged the relevant tactics to create the mate-retention categories defined by Buss (1988). Alpha reliabilities for the tactics and categories were acceptable, ranging from $\alpha = 0.50$ to 0.87, with a mean of 0.71 (see Table 4.1). We then correlated women's reports of their partners' scores on each of the mate-retention categories with women's reports of their partners' scores on the CBI, VAI, and IAI.

The results supported predictions 1.1, 1.2, and 1.3: women's reports of their partners' use of direct guarding correlated positively with their reports of their partners' scores on the CBI, VAI, and IAI; r (471) = 0.68, 0.45, and 0.37, respectively (all P values < 0.014). The results also supported predictions 2.1, 2.2, and 2.3: women's reports of their partners' use of intersexual negative inducements correlated positively with their reports of their partners' scores on the CBI, VAI, and IAI; r (471) = 0.62, 0.33, and 0.26, respectively (values all P values < 0.014). The results did not support prediction 3: women's reports of their partners' use of positive inducements did not correlate negatively with their reports of their partners' scores on the CBI, VAI, or IAI; r (471) = 0.32 (P < 0.014), 0.16, (P < 0.014), and 0.10 (not significant), respectively. Consistent with prediction 4.1, women's reports of their partners' use of public signals of possession correlated positively with their reports of their partners' scores on the CBI; r (471) = 0.26 (P < 0.014). Predictions 4.2 and 4.3 failed to receive support: women's reports of their partners' use of public signals of possession correlated positively but not significantly with their reports of their partners' scores on the VAI and IAI; r (471) = 0.09 and 0.08, respectively (both P values > 0.014).

As in study 1, we wanted to identify which tactics and acts best predicted (numerically) the occurrence and consequences of violence in mateships. Following the procedure used in study 1, we standardized separately with unit-weighting scores on the VAI and IAI and then averaged these standardized scores into a composite OVI; $\alpha = 0.91$. We first correlated scores on each of the mate-retention tactics with scores on the OVI. These correlations are shown in the far right column of Table 4.2. Concealment of mate showed the highest-ranking correlation with scores on the OVI, followed by emotional manipulation, vigilance, monopolization of time, and punish mate's infidelity threat. Love and care showed the lowest-ranking correlation with scores on the OVI, followed by verbal possession signals and resource display. As in study 1, the relationships between women's reports of men's mate retention and scores on the VAI and IAI paralleled those with the OVI and, therefore, are not discussed further but are displayed in the middle two columns of Table 4.2. For reportorial completeness, Table 4.2 presents the correlations between women's reports of men's mate retention and scores on the VAI, IAI, CBI, and OVI.

To identify the specific mate-retention acts that were the best predictors of violence, we computed correlations between each of the mate-retention acts and scores on the OVI. These act-level analyses revealed that 63 of the 104

Table 4.2. *Study 2: correlations between women's reports of men's mate retention and scores on the CBI, VAI, IAI, and OVI.*

Mate-retention category (α)/ mate-retention tactic (α)	CBI	VAI	IAI	OVI	Rank
Direct guarding (0.83)	0.68*	0.45*	0.37*	0.45*	
Vigilance (0.83)	0.59*	0.38*	0.31*	0.38*	3
Concealment of mate (0.68)	0.56*	0.43*	0.40*	0.46*	1
Monopolization of time (0.81)	0.62*	0.36*	0.25*	0.35*	4
Intersexual negative inducements (0.81)	0.62*	0.33*	0.26*	0.33*	
Jealousy induction (0.72)	0.36*	0.19*	0.12	0.19*	7.5
Punish mate's infidelity threat (0.74)	0.59*	0.30*	0.24*	0.31*	5
Emotional manipulation (0.86)	0.61*	0.40*	0.40*	0.43*	2
Commitment manipulation (0.50)	0.30*	0.16*	0.09	0.14*	10
Derogation of competitors (0.79)	0.49*	0.21*	0.12	0.19*	7.5
Positive inducements (0.81)	0.32*	0.16*	0.10	0.14*	
Resource display (0.87)	0.12	0.06	0.03	0.05	14
Sexual inducements (0.65)	0.36*	0.20*	0.12	0.17*	9
Appearance enhancement (0.83)	0.20*	0.10	0.05	0.08	12.5
Love and care (0.73)	0.10	0.02	0.01	0.01	16
Submission and debasement (0.78)	0.42*	0.22*	0.16*	0.21*	6
Public signals of possession (0.81)	0.26*	0.09	0.08	0.10	
Verbal possession signals (0.73)	0.20*	0.02	0.02	0.03	15
Physical possession signals (0.81)	0.19*	0.07	0.08	0.08	12.5
Possessive ornamentation (0.70)	0.28*	0.14*	0.11	0.13*	11

Note: $N = 471$; α = alpha reliability. Rank is the rank order of the magnitude of the correlation between the mate-retention tactic and scores on the OVI.
*$P < 0.014$.

mate-retention acts correlated significantly and positively with scores on the OVI (these correlations are available upon request). The acts "Did not let me talk to others of the opposite sex," "Cried when I said I might go out with someone else," "Cried in order to keep me with him," "Threatened to harm himself if I ever left," and "Read my personal mail" were the five highest-ranking correlations (r values $= 0.44$, 0.40, 0.39, 0.37, and 0.36, respectively; all P values < 0.01). Three of these acts are included within the tactic emotional manipulation, and accordingly, emotional manipulation was the second-highest tactic-level predictor of violence.

According to women's reports of their partners' behaviors, use of direct guarding and intersexual negative inducements is related positively to female-directed violence. Contrary to expectation, the use of positive inducements also is related positively to female-directed violence. Women who reported that their partners more frequently use the mate-retention tactics concealment of mate, emotional manipulation, vigilance, monopolization of time, and punish mate's infidelity threat reported the most violence in their relationships. The same pattern of findings emerged when we controlled for the woman's age, her partner's age, and the length of their relationship (analyses available upon request).

COMPARING THE RESULTS FOR MEN'S SELF-REPORTS (STUDY 1) AND WOMEN'S PARTNER-REPORTS (STUDY 2)

Table 4.3 presents descriptive statistics for the target variables for men's and women's reports and displays the results of tests for sex differences along these target variables. Women in study 1 (i.e. the surveyed men's partners) were older than the women in study 2 (i.e. the surveyed women), but men from the two studies did not differ in age. The length of the relationship reported by men in study 1 was longer than the length of the relationship reported by women in study 2. Reports of men's use of direct guarding, public signals of possession, and scores on the VAI, IAI, and OVI did not differ significantly between the two samples.

Relative to women's reports of their partners' behavior, men self-reported more frequent use of intersexual negative inducements, positive inducements, and controlling behavior. Although not anticipated, the sex difference in reported frequency of controlling behaviors is not surprising upon examination of the acts included in the CBI. More than half of the acts do not require the woman's physical presence or knowledge, for example "Deliberately keep her short of money" and "Check her movements." In addition, such acts might be more effective if the woman is not aware of their occurrence. The discrepancy between men and women's reports of men's intersexual negative inducements and positive inducements merits further investigation.

Comparing the correlations obtained from men's reports (study 1) to those obtained from women's reports (study 2) reveals that the sexes provide corroborative reports about which tactics numerically best predicted violence. Only the correlations between the mate-retention tactics and the OVI are discussed here (other correlation comparisons are available upon request). Spearman's rank order correlation indicates a strong positive relationship between (a) the ranks of the correlations between men's reports of their performance of mate-retention tactics and female-directed violence in study 1 (far-right column of Table 4.1) and (b) the ranks of the correlations between women's reports of their partner's performance of mate-retention tactics and female-directed violence in study 2 (far-right column of Table 4.3); r_s (14) = 0.76 ($p < 0.01$).

Table 4.3. *Descriptive statistics for target variables in studies 1 and 2.*

Target variable	Men's reports (study 1)		Women's reports (study 2)		
	Mean	SD	Mean	SD	t
Man's age (years)	24.2	7.9	23.7	6.6	0.99
Woman's age (years)	23.0	7.3	21.5	5.4	3.83*
Length of relationship (months)	37.2	59.7	28.7	38.4	2.74*
Men's direct guarding[a]	2.5	2.5	2.1	2.5	2.05
Men's intersexual negative inducements[b]	2.1	2.2	1.6	2.0	3.75*
Men's positive inducements[c]	6.5	2.6	5.7	2.8	4.35*
Men's public signals of possession[d]	5.5	2.5	5.7	3.2	−0.85
Men's CBI score[ef]	9.7	9.2	6.5	8.9	5.31*
Men's VAI score[f]	3.3	5.5	3.3	7.6	0.07
Men's IAI score[gh]	0.3	1.0	0.5	1.8	−1.83
Men's OVI score[h]	1.8	3.0	2.0	4.8	−0.68

Note: For study 1, $N = 461$; for study 2, $N = 560$. The t values were produced by independent means tests.

[a] Composite variable (see text), ranging from 0 (least frequent use of direct guarding) to 54 (most frequent use).

[b] Composite variable (see text), ranging from 0 (least frequent use of intersexual negative inducements) to 84 (most frequent use).

[c] Composite variable (see text), ranging from 0 (least frequent use of positive inducements) to 78 (most frequent use).

[d] Composite variable (see text), ranging from 0 (least frequent use of public signals of possession) to 45 (most frequent use).

[e] CBI (see text), ranging from 0 (minimum use of controlling behaviors) to 105 (maximum use).

[f] VAI (see text), ranging from 0 (minimum use of violence) to 125 (maximum use).

[g] IAI (see text), ranging from 0 (minimum occurrence of injuries) to 95 (maximum occurrence).

[h] Composite variable (see text), average of the VAI and IAI.

*$P < 0.01$ (two-tailed).

Men in Study 1 did not consistently underreport the violence they inflicted on their partners, relative to women's reports provided in study 2. The men that participated in study 1, however, were not partnered to the women that participated in study 2. Previous literature on discrepancy in reports is based on comparisons of the reports of partnered men and women (e.g. Dobash *et al.*, 1998; Magdol *et al.*, 1997). The use of independent samples of men and women in the current studies makes difficult a direct comparison with previous literature.

Study 1 secured men's reports of their mate retention and violence in romantic relationships. Many of the correlations between the use of mate retention and violence were statistically significant but small in magnitude. Study 2 secured women's reports of their partners' mate retention and violence. The correlations identified in Study 2 between men's use of mate retention and violence were generally larger numerically than those identified in study 1. Using women's reports of their partners' mate retention may be problematic, however, because men may be in a better position to report on their own mate-retention behaviors, some of which occur outside the awareness of their partner (e.g. "He had his friends check up on her"). Because women report relationship violence with relative accuracy and men may be better able to report accurately their use of mate-retention behaviors, we conducted a third study to secure these reports in a sample of married couples. Married couples served as participants for study 3. Husbands reported their use of mate-retention behaviors and their wives reported husbands' use of violence.

Appendix 4.2   Study 3: husbands' reports of their mate retention and wives' reports of husbands' violence

In study 3, we collected husbands' reports of their mate-retention behaviors and wives' reports of their husbands' violence. Using these data, we tested four predictions paralleling those tested in studies 1 and 2 (Appendix 4.1).

METHODS

Participants

Participants were 214 individuals, 107 men and 107 women, who had been married less than 1 year. Participants were obtained from the public records of marriage licenses issued within a large county in midwestern USA. All couples married within the designated time period were invited to participate in this study. The mean age of husbands was 25.5 years (SD = 6.6 years). The mean age of wives was 24.8 years (SD = 6.2 years). Additional details about this sample can be found in Buss (1992).

Materials

Husbands completed the MRI (Buss, 1988). Wives completed the Spouse Influence Report (SIR; Buss, 1992; Buss *et al.*, 1987), which is designed to assess behaviors that husbands use to influence or manipulate their partners. Items ranged from nonviolent manipulative behaviors to violent manipulative behaviors. Example items include "He tells me how happy he'll be if I do it," and "He yells at me so I'll do it." Responses are recorded on a seven-point Likert-type scale anchored by 1 (not at all likely to do this) and 7 (extremely likely to do this).

Procedure

Participants engaged in two separate episodes of assessment. First, they received through the mail a battery of instruments to be completed at home. Husbands completed the MRI and other measures designed for different studies. Second, participants came to a testing session 1 week after receiving the first battery. Spouses were separated to preserve independence and to prevent contamination due to discussion. During this session, wives' completed the SIR and other measures designed for different studies.

RESULTS AND DISCUSSION

As in studies 1 and 2, we standardized with unit weighting the mate-retention tactics and then averaged the relevant tactics to create the mate-retention categories defined by Buss (1988). Alpha reliabilities for the tactics and categories were acceptable, ranging from $\alpha = 0.46$ to 0.83, with a mean of 0.67 (see Table 4.4). The female-directed violence variable used in study 3 differed from that used in studies 1 and 2. Study 3 did not include the CBI, VAI, or IAI. To measure violence in study 3, we standardized with unit weighting and then averaged responses to one act from the MRI ("He hit me when he caught me flirting with someone else," which was excluded from other analyses of mate retention) with two acts from the SIR ("He hit me so I will do it," "He implied the possibility of physical harm if I didn't do"). Responses to these three acts produced a reliable index of wives' reports of their husbands' violence; $\alpha = 0.70$ (mean inter-item correlation, r [105] $= 0.43$; the results do not change when we exclude the SRI item in which violence is implied rather than committed; analyses available upon request).

We then correlated husbands' reports of their mate retention with wives' reports of violence. Consistent with predictions 1, 2, and 4, husbands' self-reported use of direct guarding, intersexual negative inducements, and public signals of possession were related positively to wives' reports of husbands' violence; r (105) $= 0.43$, 0.41, and 0.32 (all P values < 0.014). Prediction 3 was not supported: husbands' use of positive inducements was not related negatively to wives' reports of husbands' violence; r (105) $= 0.23$ (not significant).

As in studies 1 and 2, we wanted to identify which of the mate-retention tactics and acts best predicted (numerically) violence against women. We correlated scores on each of the tactics with violence against wives. These correlations are shown in Table 4.4. Vigilance showed the highest-ranking correlation with violence against wives, followed by emotional manipulation,

Table 4.4. *Study 3: correlations between husbands' self-reported mate retention and their wives' reports of violence.*

Mate retention category (α)/mate-retention tactic (α)	Relationship violence	Rank
Direct guarding (0.76)	0.43*	
Vigilance (0.74)	0.50*	1
Concealment of mate (0.67)	0.18	11
Monopolization of time (0.78)	0.36*	3
Intersexual negative inducements (0.73)	0.41*	
Jealousy induction (0.68)	0.17	12
Punish mate's infidelity threat (0.82)	0.34*	6
Emotional manipulation (0.75)	0.43*	2
Commitment manipulation (0.46)	0.19	10
Derogation of competitors (0.67)	0.34*	6
Positive inducements (0.71)	0.23	
Resource display (0.67)	0.12	13.5
Sexual inducements (0.66)	0.31*	9
Appearance enhancement (0.66)	0.04	15
Love and care (0.66)	−0.03	16
Submission and debasement (0.61)	0.32*	8
Public signals of possession (0.78)	0.32*	
Verbal possession signals (0.49)	0.34*	6
Physical possession signals (0.71)	0.12	13.5
Possessive ornamentation (0.67)	0.35*	4

Note: $N = 107$; $\alpha =$ alpha reliability. Rank is the rank order of the magnitude of the correlation between the mate-retention tactic and scores on the relationship violence index.
*$P < 0.014$.

monopolization of time, and possessive ornamentation. Love and care showed the lowest-ranking correlation with violence against wives, followed by appearance enhancement.

To identify the specific mate-retention acts that were the best predictors of violence, we computed correlations between each of the mate-retention acts and the relationship violence score. These act-level analyses revealed that 38 of the 104 mate-retention acts correlated significantly and positively with relationship violence (these correlations are available upon request). The acts "Told my partner that someone of my same sex was out to use my partner," "Hung up a picture of my partner so that others would know my partner was taken," "Dropped by unexpectedly to see what my partner was doing," "Told my partner that I would 'die' if my partner ever left," and "Called to make sure my partner was where she said she would be" were the five highest-ranking correlations

(r values $= 0.50$, 0.46, 0.44, 0.40, and 0.40, respectively, all P values < 0.01). Two of these five acts are included in the tactic vigilance and, accordingly, vigilance was the best tactic-level predictor of violence in study 3.

COMPARING THE RESULTS OF STUDY 3 WITH THE RESULTS OF STUDIES 1 AND 2

Comparing the correlations between men's mate retention and female-directed violence obtained from men's reports (study 1) with those obtained from husbands' and their wives' reports (study 3) reveals that, of the study comparisons, these two perspectives were in least agreement on which tactics best predicted violence. Correlations between violence against women and men's use of emotional manipulation and monopolization of time, however, were among the highest-ranking correlations in both studies (see Tables 4.1 and 4.4). Emotional manipulation produced the highest-ranking correlation in study 1 and the second-highest ranking correlation in study 3, and monopolization of time produced the third-highest ranking correlation in both studies 1 and 3. Only the correlations between the mate-retention tactics and the measures of violence are discussed here (other correlation comparisons are available upon request). Spearman's rank order correlation revealed a positive but not statistically significant relationship between the ranks of the correlations of female-directed violence (as assessed by the OVI) with the mate-retention tactics in study 1 and the ranks of the correlations of female-directed violence with these tactics in study 3; r_s (14) $= 0.39$ (not significant). Some of the discrepancy between the two studies about which tactics best predicted violence might be attributable to the fact that the measures of violence differed in studies 1 and 3. The use of identical measures of violence may have reduced this discrepancy.

Comparing the correlations obtained from women's reports (study 2) with those obtained from husbands' reports and their wives' reports (study 3) revealed some agreement on which tactics best predicted violence in mateships. Spearman's rank order correlation indicated a positive and statistically significant relationship between the ranks of the correlations between the mate-retention tactics with female-directed violence (as assessed by the OVI) in study 2 and the ranks of the correlations between the mate-retention tactics with female-directed violence in study 3; r_s (14) $= 0.60$, ($P < 0.01$). As noted for comparisons of the results of studies 1 and 3, some of the discrepancy between studies 2 and 3 on which tactics best predicted violence in mateships could be attributable to the fact that the measures of violence differed across the two studies. In the general discussion section of this chapter we summarize the key findings generated from these three studies.

References

Buss, D. M. (1988). From vigilance to violence: tactics of mate retention in American undergraduates. *Ethology and Sociobiology*, **9**, 291–317.

Buss, D. M. (1992). Manipulation in close relationships: five personality factors in interactional context. *Journal of Personality*, **60**, 477–99.

Buss, D. M. (2000). *The Dangerous Passion*. New York: Free Press.

Buss, D. M. and Craik, K. H. (1983). The act frequency approach to personality. *Psychological Review*, **90**, 105–26.

Buss, D. M. and Shackelford, T. K. (1997). From vigilance to violence: mate retention tactics in married couples. *Journal of Personality and Social Psychology*, **72**, 346–61.

Buss, D. M., Gomes, M., Higgins, D. S., and Lauterbach, K. (1987). Tactics of manipulation. *Journal of Personality & Social Psychology*, **52**, 1219–29.

Buss, D. M., Larsen, R. J., Westen, D., and Semmelroth, J. (1992). Sex differences in jealousy: evolution, physiology and psychology. *Psychological Science*, **3**, 251–55.

Cohen, J. and Cohen, P. (1983). *Applied Multiple Regression/Correlation Analysis for the Behavioral Sciences*, 2nd edn. Hillsdale, NJ: Lawrence Erlbaum Associates.

Daly, M. and Wilson, M. (1988). *Homicide*. Hawthorne, NY: Aldine de Gruyter.

Daly, M., Wilson, M., and Weghorst, J. (1982). Male sexual jealousy. *Ethology and Sociobiology*, **3**, 11–27.

Dobash, R. E., Dobash, R. P., Cavanagh, K., and Lewis, R. (1995). Evaluating criminal justice programmes for violent men. In R. E. Dobash, R. P. Dobash, and L. Noaks, eds., *Gender and Crime*. Cardiff: University of Wales Press, pp. 358–89.

Dobash, R. E., Dobash, R. P., Cavanagh, K., and Lewis, R. (1996). *Research Evaluation of Programmes for Violent Men*. Edinburgh: Scottish Office Central Research Unit.

Dobash, R. E., Dobash, R. P., Cavanagh, K., and Lewis, R. (1998). Separate and intersecting realities: a comparison of men's and women's accounts of violence against women. *Violence Against Women*, **4**, 382–414.

Dutton, D. G. (1998). *The Abusive Personality*. New York: Guilford Press.

Gangestad, S. W., Thornhill, R., and Garver, C. E. (2002). Changes in women's sexual interests and their partners' mate-retention tactics across the menstrual cycle: evidence for shifting conflicts of interest. *Proceedings of the Royal Society of London B*, **269**, 975–82.

Goetz, A. T., Shackelford, T. K., Weekes-Shackelford, V. A. *et al.* (2005). Mate retention, semen displacement, and human sperm competition: a preliminary investigation of tactics to prevent and correct female infidelity. *Personality and Individual Difference*, **38**, 749–63.

Hays, W. L. (1988). *Statistics*, 4th edn. Fort Worth, TX: Holt, Rinehart, & Winston.

Magdol, L., Moffitt, T. E., Caspi, A., Newman, D. L., Fagan, J., and Silva, P. A. (1997). Gender differences in partner violence in a birth cohort of 21-year-olds: bridging the gap between clinical and epidemiological approaches. *Journal of Consulting and Clinical Psychology*, **65**, 68–78.

Parker, G. A. (1970). Sperm competition and its evolutionary consequences in the insects. *Biological Reviews*, **45**, 525–67.

Schmitt, D. P. and Buss, D. M. (2001). Human mate poaching: tactics and temptations for infiltrating existing mateships. *Journal of Personality and Social Psychology*, **80**, 894–917.

Shackelford, T. K. and Buss, D. M. (1997). Cues to infidelity. *Personality and Social Psychology Bulletin*, **23**, 1034–45.

Shackelford, T. K. and Buss, D. M. (2000). Marital satisfaction and spousal cost-infliction. *Personality and Individual Differences*, **28**, 917–28.

Shackelford, T. K., Goetz, A. T., Buss, D. M., Euler, H. A., and Hoier, S. (2005). When we hurt the ones we love: Predicting violence against women from men's mate retention tactics. *Personal Relationships*, **12**, 447–63.

Symons, D. (1979). *The Evolution of Human Sexuality*. New York: Oxford University Press.

Walker, L. E. (2000). *The Battered Woman Syndrome*, 2nd ed. New York: Springer Publishing Company.

Wilson, M. and Daly, M. (1992). The man who mistook his wife for a chattel. In J. Barkow, L. Cosmides, and J. Tooby eds., *The Adapted Mind*. New York: Oxford University Press, pp. 289–322.

Wilson, M., Johnson, H., and Daly M. (1995). Lethal and nonlethal violence against wives. *Canadian Journal of Criminology*, **37**, 331–61.

5

Sexual coercion and forced in-pair copulation as anti-cuckoldry tactics in humans

AARON T. GOETZ AND TODD K. SHACKELFORD
Florida Atlantic University

Introduction

Rape in humans may or may not be generated by specialized psychological adaptation (Alexander & Noonan, 1979; Palmer, 1991; Thornhill & Palmer, 2000; Thornhill & Thornhill, 1992). Although several hypotheses have been proposed, there are only two likely candidates for evolutionary explanations of rape in humans. One hypothesis posits that rape is generated by an adaptation that functions as a facultative male reproductive tactic that contributes directly to reproductive success by increasing sexual partner number (e.g. Shields & Shields, 1983; Thornhill & Thornhill, 1983). The other hypothesis posits that rape was not directly selected for over evolutionary history, but instead is a byproduct of other male psychological adaptations, particularly those associated with sexual variety and aggression (Palmer, 1991; Thornhill & Palmer, 2000).

Although the debate continues about whether human rape is generated by specialized adaptation or is generated as a byproduct, a special case of human rape presents an equally interesting question. If human rape is either due to selection pressures to increase sexual partner number, or due to other psychological adaptations, such as those associated with obtaining numerous sexual partners, then why do men in committed sexual relationships sometimes rape their partners? Researchers estimate that between 10 and 17% of women experience rape in marriage (Finkelhor & Yllo, 1985; Painter & Farrington, 1999; Russell, 1982). Moreover, particular subgroups of women may be especially at risk of experiencing rape in their marriage: 23–50% of physically abused women experience rape by their husbands (Bowker, 1983; Campbell,

1989; Frieze, 1983; Pagelow, 1981; Shields & Hanneke, 1983). The not uncommon occurrence of rape by an intimate partner poses an interesting evolutionary question, given that men in committed sexual relationships already have sexual access to their partners and thus will not increase sexual partner number by raping them.

Although sometimes referred to as *marital rape*, *spouse rape*, or *wife rape*, we use the term *forced in-pair copulation* (FIPC) to refer to the forceful act of sexual intercourse by a man against his partner's will. Before considering the case of FIPC in humans, we review briefly the animal literature on FIPC. Examining the adaptive problems and resultant evolved solutions to these problems in non-human animals may provide insight into the adaptive problems and evolved solutions in humans (and vice versa). Shackelford and LeBlanc (2001), for example, argued that because humans share with some avian species a similar mating system (social monogamy) and similar adaptive problems (e.g. paternity uncertainty in males, mate retention, cuckoldry), humans and some birds may share similar solutions to these adaptive problems. Identifying the contexts and circumstances in which FIPC occurs in non-human species may help us to understand why FIPC occurs in humans.

FIPC in non-human animals

Instances of FIPC are relatively rare in the animal kingdom, primarily because males and females of most species (over 95%) do not form long-term pair-bonds (Andersson, 1994). Without the formation of a pair-bond, FIPC, by definition, cannot occur. Many avian species form long-term pair-bonds, and researchers have documented FIPC in several of these species (Bailey, Seymour, & Stewart, 1978; Barash, 1977; Birkhead, Hunter, & Pellatt, 1989; Cheng, Burns, & McKinney, 1983; Goodwin, 1955; McKinney, Cheng, & Bruggers, 1984; McKinney & Stolen, 1982). FIPC is not performed randomly, however. FIPC reliably occurs immediately after extra-pair copulations, intrusions by rival males, and female absence in many species of waterfowl (e.g. Bailey *et al.*, 1978; Barash, 1977; Cheng *et al.*, 1983; McKinney, Derrickson, & Mineau, 1983; McKinney & Stolen, 1982; Seymour & Titman, 1979) and other avian species (e.g. Birkhead *et al.*, 1989; Goodwin, 1955; Valera, Hoi, & Kristin, 2003). FIPC following observed or suspected extra-pair copulation in these avian species is often interpreted as a sperm-competition tactic (Barash, 1977; Cheng *et al.*, 1983; Lalumière, *et al.*, 2005; McKinney *et al.*, 1984).

Sperm competition is a form of male–male postcopulatory competition. Sperm competition occurs when the sperm of two or more males simultaneously occupy the reproductive tract of a female and compete to fertilize her egg (Parker, 1970). Males can compete for mates, but if two or more males have copulated with a female within a sufficiently short period of time, males must compete for fertilizations. Thus, the observation in many avian species that FIPC immediately follows extra-pair copulations was understood as a sperm-competition tactic because the in-pair male's FIPC functioned to place his sperm in competition with sperm from an extra-pair male (Birkhead *et al.*, 1989; Cheng *et al.*, 1983). Reports of FIPC in non-human species are theoretically beneficial in that they make it difficult to claim that a male raped his partner because he wanted to humiliate, punish, or control her – as is often argued by some social scientists who study rape in humans (e.g. Pagelow, 1988).

Mounting evidence suggests that sperm competition has been a recurrent and important feature of human evolutionary history. Psychological, behavioral, physiological, anatomical, and genetic evidence reveals that ancestral women sometimes mated with multiple men within sufficiently short time periods so that sperm from two or more males simultaneously occupied the reproductive tract of one woman (Baker & Bellis, 1993; Gallup *et al.*, 2003; Goetz *et al.*, 2005; Pound, 2002; Shackelford *et al.*, 2004; Shackelford, Pound, & Goetz, 2005c; Shackelford *et al.*, 2002; Smith, 1984; Wyckoff, Wang, & Wu, 2000). This adaptive problem led to the evolution of adaptive solutions to sperm competition. For example, men display copulatory urgency, perform semen-displacing behaviors, and adjust their ejaculates to include more sperm when the likelihood of female infidelity is high (Baker & Bellis, 1993; Gallup *et al.*, 2003; Goetz *et al.*, 2005; Shackelford *et al.*, 2002).

FIPC in humans

Noting these instances of FIPC followed by extra-pair copulations in waterfowl and documentation that FIPC in humans often followed accusations of female infidelity (e.g. Finkelhor & Yllo, 1985; Russell, 1982), Wilson and Daly (1992) suggested, in a footnote, that "sexual insistence" in the context of a relationship might act as a sperm-competition tactic in humans as well. Sexual coercion in response to cues of his partner's sexual infidelity might function to introduce a male's sperm into his partner's reproductive tract at a time when there is a high risk of cuckoldry.

Thornhill and Thornhill (1992) also hypothesized that FIPC may be an anti-cuckoldry tactic designed by sperm competition. Thornhill and Thornhill argued that a woman who resists or avoids copulating with her partner might thereby be signaling to him that she has been sexually unfaithful and that the FIPC functions to decrease his paternity uncertainty (see also Gallup and

Burch, Chapter 7 of this volume). Thornhill and Thornhill argued that the fact that the rape of a woman by her partner is more likely to occur during or after a breakup – times in which men express great concern about female sexual infidelity – provides preliminary support for the hypothesis. Finkelhor and Yllo (1985), for example, found that over two-thirds of the women in their sample were raped by their partners at the end of the relationship, whereas only 31% were raped early in the relationship and 40% were raped in the middle of the relationship. Thornhill and Thornhill also cited research by Frieze (1983) indicating that women who were physically abused and raped by their husbands rated them to be more sexually jealous than did women who were abused but not raped. Similar arguments were presented by Thornhill and Palmer (2000), and Lalumière *et al.* (2005) suggested that antisocial men who suspect that their female partner has been sexually unfaithful may be motivated to engage in FIPC.

FIPC and sexual coercion in intimate relationships

FIPC is just one aspect of a constellation of behaviors that comprise sexual coercion in intimate relationships (Koss & Oros, 1982; Shackelford & Goetz, 2004; Weis & Borges, 1973), but the explicit use of force to obtain sexual intercourse is likely the most costly, to both the victims and perpetrators. The severity of FIPC is demonstrated in the finding that physically abused women who experience FIPC have significantly more negative health symptoms and gynecological problems than women who are physically abused but not raped by their partners (e.g. Campbell & Soeken, 1999). Moreover, FIPC may be more traumatizing than forced copulation by a stranger (Bart, 1975; Russell, 1982). Given the potentially devastating costs associated with FIPC, sexual coercion is likely to take more subtle forms. Shackelford and Goetz (2004), for example, documented that men sexually coerce their partners by hinting about withholding benefits, threatening defection, and manipulating their partner's commitment to the relationship (e.g. "If you love me, you'll have sex with me"). By using more discreet forms of sexual coercion (as opposed to using explicit force), men may avoid inflicting on their partners the costs associated with FIPC and they may avoid their partner's defection from the relationship.

Given this reasoning, subtle forms of sexual coercion in the context of an intimate relationship are likely more prevalent than more explicit forms such as FIPC. Therefore, not only may FIPC function as a sperm-competition tactic, but all forms of sexual coercion in the context of an intimate relationship may be a product of sperm competition. This leads to the first hypothesis.

> **Hypothesis 1:** Men's sexual coercion in the context of an intimate relationship is related positively to his partner's infidelities.

Sexual coercion and mate retention

Research has documented that men engage in an assortment of behaviors designed to prevent their partner's infidelity. Using an act-nomination procedure, Buss (1988) identified specific "mate-retention" behaviors that men use to guard or to retain their mates. Subsequent research has found that men increase their mate-retention behaviors when their partner is of greater reproductive value (as indexed by her age and physical attractiveness), when she is more likely to engage in extra-pair copulations, and when she is near ovulation (Buss & Shackelford, 1997; Gangestad, Thornhill, & Garver, 2002; Goetz *et al.*, 2005). Moreover, research has documented that men use mate-retention behaviors in conjunction with, and not alternatively to, other anti-cuckoldry tactics (e.g. Goetz *et al.*, 2005; Shackelford, Goetz, & Buss, 2005a). In other words, men who perceive that their likelihood of being cuckolded is high use an arsenal of anti-cuckoldry tactics to guard their paternity. If sexual coercion is a sperm-competition tactic designed to "correct" a partner's sexual infidelity, then men who sexually coerce their partners also should perform more mate-retention behaviors. This leads to the second hypothesis.

> **Hypothesis 2:** Men's sexual coercion in the context of an intimate relationship is related positively to their mate-retention behaviors.

It may be that a proportion of sexually coercive behaviors (in the context of an intimate relationship) are performed by antisocial men who aim to punish, humiliate, or control their partners *independent of their perception of cuckoldry risk*. We are not arguing that all sexual coercion and FIPCs are the output of evolved psychological mechanisms designed to reduce the risk of being cuckolded. Instead, we are suggesting that a significant amount of sexual coercion might be the result of male-evolved psychology associated with guarding their paternity.

A secondary goal of this research was to obtain from a relatively young sample of adults the prevalence estimates of FIPC. Previous studies assessing the prevalence of FIPC have only assessed FIPC in marriages and have restricted their sample to women (Finkelhor & Yllo, 1985; Painter & Farrington, 1999; Russell, 1982). This research will contribute uniquely to the existing literature by securing prevalence estimates for FIPC from men and women who are in a committed relationship (for a minimum of 1 year) and not necessarily married.

This chapter highlights some of our recent research (see Goetz & Shackelford, 2006), in which we investigated men's sexual coercion, their risk of sperm competition, and their mate-retention behaviors in two studies. Study 1 (Appendix 1) focused on men's reports of their own sexual coercion in the current relationship, their perception of their partners' infidelities, and their own mate-retention behaviors. Study 2 (Appendix 2) focused on women's reports of their partners' sexual coercion in the current relationship, their own infidelities, and their partners' mate-retention behaviors.

Comparisons between men's self-reports (study 1) and women's partner-reports (study 2)

Because research indicates that men's reports of their sexual coercion and mate retention may be less reliable or less accurate than women's reports of their partners' coercive behaviors and mate-retention behaviors (Dobash *et al.*, 1998; Magdol *et al.*, 1997), it may be appropriate to place greater weight on women's reports. We performed Fisher's r-to-z transformations to compare the magnitude of correlations generated by men's self-reports (study 1; Appendix 5.1) and women's partner-reports (Study 2; Appendix 5.2). For hypothesis 1, the correlation obtained from the men's data ($r = 0.25$) was not significantly different than the correlation obtained from the women's data ($r = 0.32$); $z = -0.86$ (*not significant*). The magnitudes of the relationship between men's sexual coercion and his partner's infidelities were not significantly different between the samples. For hypothesis 2, the correlation obtained from the men's data ($r = 0.15$) was significantly lower than the correlation obtained from the women's data ($r = 0.34$); $z = -2.30$ ($P < 0.05$). The magnitude of the relationship between men's sexual coercion and mate retention was significantly greater for women's partner-reports than for men's self-reports.

Next, we performed Fisher's r-to-z transformations to identify differences between the correlations of sexual coercion with each of the 19 mate-retention tactics for the data provided by men (correlations in the first column of Table 5.1; see Appendix 5.1) and the correlations of sexual coercion with each of the 19 mate-retention tactics for the data provided by women (correlations in the second column of Table 5.1). We identified significant differences in correlations for nine mate-retention tactics: vigilance, concealment of mate, monopolization of time, emotional manipulation, commitment manipulation, resource display, sexual inducements, submission and debasement, and possessive ornamentation. For each of these tactics, correlations of sexual coercion with mate retention for the self-report data provided by men were significantly lower than correlations of sexual coercion with mate retention for the partner-report data provided by women.

Finally, we tested the difference between the prevalence estimates of FIPC generated by men's self-reports and women's partner-reports. The prevalence of FIPC obtained from the men's self-report data (7.3%) was not significantly different than the prevalence of FIPC obtained from the women's partner-report data (9.1%); $z = -0.75$ (*not significant*). Although the prevalence of FIPC is numerically greater according to women's partner-reports, the percentages were not significantly different between the samples.

General discussion

FIPC has been documented in several avian species and reliably occurs immediately after an observed or suspected extra-pair copulation (see Lalumière *et al.*, 2005, for a review). This behavior has been interpreted as a sperm-competition tactic because the in-pair male's FIPC functions to place his sperm in competition with sperm from an extra-pair male (Birkhead *et al.*, 1989; Cheng *et al.*, 1983). FIPC is not unique to birds; researchers estimate that between 10 and 17% of women experience FIPC in marriage (Finkelhor & Yllo, 1985; Painter & Farrington, 1999; Russell, 1982). Narratives provided by victims of FIPC reveal that a remarkable proportion of FIPCs follow accusations of female infidelity and suggest a strong and reliable link between male sexual jealousy and the occurrence of FIPC (see Bergen, 1996; Finkelhor & Yllo, 1985; Frieze, 1983; Russell, 1982; Walker, 1979). Consequently, several researchers have interpreted FIPC in humans as a sperm-competition tactic (Lalumière *et al.*, 2005; Thornhill & Thornhill, 1992; Wilson & Daly, 1992).

The current studies tested specific hypotheses derived from the general hypothesis that sexual coercion in the context of an intimate relationship may function as a sperm-competition tactic. We hypothesized that men's sexual coercion in the context of an intimate relationship is related positively to his partner's infidelities and that men's sexual coercion is related positively to their mate-retention behaviors. The results from study 1 and study 2 (see Appendices 5.1 and 5.2) supported the hypotheses. According to men self-reports and women's partner-reports, men who used more sexual coercion in their relationship are mated to women who had been or were likely to be unfaithful, and these men also are likely to use more mate-retention behaviors.

Although the correlations of the tests of the hypotheses and the prevalence of FIPC were only significantly different between the sexes in the test of hypothesis 2 (i.e. the relationship between sexual coercion and mate retention), the correlations and prevalence of FIPC were numerically greater for women's partner-report data. Because women's reports are likely to be more accurate and more reliable than men's reports (Dobash *et al.*, 1998; Magdol *et al.*, 1997), it may be appropriate to place greater weight on women's reports of men's sexual coercion and mate retention.

One limitation of the current research is in its design. We present correlational analyses that prevent strong statements about causal relationships. We speculate that women's infidelities cause men to use sexual coercion as a paternity guard. The data are consistent with this interpretation, but we cannot yet rule out an alternative, reverse causal relationship – that men's sexual coercion causes women to become unfaithful. A methodology that includes repeated assessments of the key variables over time, such as a daily diary study, would allow for the identification of causal relationships.

A clear future direction is to identify the environmental inputs that activate the proposed psychological mechanisms associated with sexual coercion and FIPC. Cues to a partner's sexual infidelity may be unequivocal, such as admission or observation of the infidelity, but most cues are probably more cryptic, such as apathy toward her partner, sudden decreased sexual interest in her partner, and subtle changes in her normal routine (Shackelford & Buss, 1997). It would be valuable to identify the specific cues that motivate, in some men, sexual coercion and FIPC.

Another future avenue of research could involve measuring phallometry in men convicted of FIPC. Phallometry is the research method of measuring erectile responses while presenting stories or pictures of sexual and non-sexual stimuli. Meta-analyses of phallometric studies strongly suggest that rapists respond differently to sexual stimuli than non-rapists (Hall, Shondrick, & Hirschman, 1993; Lalumière & Quinsey, 1994). If FIPC is unrelated to rape in general (e.g. stranger rape), then men convicted of FIPC should show phallometric responses that differ from men convicted of stranger rape. For example, men convicted of FIPC might have phallometric responses more like "normal" men, because the motivation underlying FIPC is different than that underlying general forced copulation.

Conclusion

Because cuckoldry poses a substantial reproductive cost for males of paternally investing species, men are expected to have evolved a host of adaptations to confront the adaptive problem of cuckoldry. One such adaptation may be a sperm-competition tactic whereby sexual coercion and FIPC function to increase the likelihood that the in-pair male, and not a rival male, sires the offspring that his partner might produce.

Appendix 5.1   Study 1: men's self-reports

METHODS

Participants

Two hundred and forty six men, each of whom was in a committed, sexual relationship with a woman for at least one year, participated in this study. About half of the participants were drawn from a university in southern Florida and the other half from surrounding communities. The mean age of the participants was 25.1 years (SD = 7.1 years), the mean age of the participants' partners was 23.8 years (SD = 6.7 years), and the mean relationship length was 46.3 months (SD = 49.1 months).

Materials

Participants completed a survey that included several sections. The first section requested demographic information, including the participant's age, his partner's age, and the length of his current relationship. The second section asked four questions to assess his partner's past sexual and emotional infidelities and her likelihood of committing a sexual and emotional infidelity in the future: "As far as you know, has your current partner had sexual inter-course with someone other than you since you have been involved in a relation-ship together?," "As far as you know, has your current partner fallen in love with someone other than you since you have been involved in a relation-ship together?," "How likely do you think it is that your current partner will in the future have sexual intercourse with someone other than you, while in a relationship with you?," and "How likely do you think it is that your current partner will in the future fall in love with someone other than you, while in a relationship with you?" Responses were recorded using a 10-point Likert-type scale ranging from 0 (definitely no/not at all likely) to 9 (definitely yes/extremely likely).

To assess men's mate-retention behaviors, the third section of the survey included the Mate Retention Inventory (MRI; Buss, 1988; Buss & Shackelford, 1997; Shackelford, Goetz, & Buss, 2005a), which asked how often the participant had performed 104 mate-retention acts in the last month, ranging from 0 (never) to 3 (often). Example acts include: "Did not let my partner talk to other men," "Held my partner's hand when other men were around," and "Introduced my partner as my spouse or romantic partner." Buss (1988) grouped these 104 acts into 19 tactics of mate retention. Previous research has established the reliabil-ity, validity, and utility of the MRI as an assessment of mate-retention behav-iors (Buss, 1988; Buss & Shackelford, 1997; Shackelford *et al.*, 2005a).

To assess men's sexual coercion in the current relationship, the last section of the survey included the male version of the Sexual Coercion in Intimate Relationships Scale (SCIRS; Shackelford & Goetz, 2004). The SCIRS asked how often the participant performed 34 sexually coercive acts in the last month. Responses were recorded using a six-point Likert-type scale with the following values: 0 = act did not occur in the past month, 1 = act occurred 1 time in the past month, 2 = act occurred 2 times in the past month, 3 = act occurred 3 to 5 times in the past month, 4 = act occurred 6 to 10 times in the past month, and 5 = act occurred 11 or more times in the past month. Items in the SCIRS vary in subtlety, ranging from hinting and subtle manipulations to outright physical force. Example items include: "I hinted that I would withhold benefits that my partner depends on if she did not have sex with me," "I told my partner that if she loved me, she would have sex with me," and "I threatened to have sex with another woman if my partner did not have sex with me." Previous research has established the reliability, validity, and utility of the SCIRS as an assessment of sexual coercion in intimate relationships (Shackelford & Goetz, 2004).

Procedure

Three criteria must have been met to qualify for participation. The prospective participant must be (1) male, (2) at least 18 years of age, and (3) currently involved in a committed, sexual relationship with a woman for at least 1 year. Upon the prospective participant's arrival at the scheduled time and location, the researcher confirmed that the prospective participant met the three participation criteria. If the criteria were met, the researcher handed the participant a consent form, the survey, and a security envelope. The participant was instructed to read and sign the consent form, complete the survey, place the completed survey in the envelope, and then seal the envelope. The participant was instructed not to seal the consent form inside the envelope to maintain anonymity. Upon completion of the survey, the researcher explained to the participant the purpose of the study, answered any questions, and thanked the participant for his participation.

RESULTS AND DISCUSSION

Prior to analyses, we created the composite variable *female infidelity* ($\alpha = 0.72$). Female infidelity is the sum of four variables: (1) partner's past sexual infidelity, (2) partner's past emotional infidelity, (3) partner's likelihood of future sexual infidelity, and (4) partner's likelihood of future emotional infidelity.

We calculated scores for men's sexual coercion using responses to the SCIRS (Shackelford & Goetz, 2004). The alpha reliability for the summed responses to the 34 items of the SCIRS was $\alpha = 0.95$. We calculated scores for 19 mate-retention tactics using the 104 items in the MRI (Buss, 1988; Shackelford et al., 2005a). The alpha reliability for the entire inventory (with responses summed across items) was $\alpha = 0.96$. Alpha reliabilities for the 19 mate-retention tactics varied in this study from $\alpha = 0.51$ to 0.90, with a mean of $\alpha = 0.75$.

Hypothesis 1 stated that men's sexual coercion in the context of an intimate relationship is related positively to their partner's infidelities. Consistent with this hypothesis, men's sexual coercion correlated positively with their perceptions of their partner's infidelities; r (244) = 0.25 ($P < 0.001$, two-tailed; all P values generated to test the hypotheses in both studies were two-tailed). Men who reported that their partners had been or were likely to be unfaithful reported using more sexual coercion with their partners.

Hypothesis 2 stated that men's sexual coercion in the context of an intimate relationship is related positively to their mate-retention behaviors. Consistent with this hypothesis, men's sexual coercion correlated positively with their mate-retention behaviors; r (244) = 0.15 ($P < 0.05$). Men who reported using more sexual coercion in their relationship reported using more mate-retention behaviors. Although the focus of hypothesis 2 was the relationship between men's sexual coercion and their broad use of mate-retention behaviors, we also report in Table 5.1 correlations between men's sexual coercion and the 19 mate-retention tactics identified by Buss (1988) and his colleagues (Buss & Shackelford, 1997; Shackelford et al., 2005a). Nine of the 19 mate-retention tactics showed significant positive correlations with men's use of sexual coercion in the current relationship: vigilance, monopolization of time, jealousy induction, punish mate's infidelity threat, emotional manipulation, derogation of competitors, sexual inducements, intrasexual threats, and violence against rivals.

A secondary goal of this study was to obtain a prevalence estimate of FIPC in a sample of young adults. We asked men if they had ever physically forced their current partner to have sex with them or if they had ever initiated sex with their current partner when she was unaware (e.g. when she was asleep, drunk, or on medication) and continued against her will. Eighteen of the 246 men in this sample (7.3%) admitted to engaging in at least one FIPC with their current partner. Previous studies generating prevalence estimates of FIPC assessed whether FIPCs occurred in the current or previous marriages (Finkelhor & Yllo, 1985; Painter & Farrington, 1999; Russell, 1982). Data presented in the current study are unique in that they were secured from a sample of relatively young men who are in a committed relationship (for a minimum of one year) and not necessarily married. The prevalence of FIPC in this study may be lower than that reported in previous studies because we sampled men, who

Table 5.1. *Correlations between men's sexual coercion and men's mate retention according to men self-reports (study 1) and women's partner-reports (study 2).*

Mate-retention tactic	Men's reports of sexual coercion	Women's reports of sexual coercion	Difference between correlations (z)
Vigilance	0.18[†]	0.35[‡]	−2.08*
Concealment of mate	0.06	0.56[‡]	−6.49[‡]
Monopolization of time	0.22[‡]	0.43[‡]	−2.68[†]
Jealousy induction	0.26[‡]	0.36[‡]	−1.26
Punish mate's infidelity threat	0.23[‡]	0.33[‡]	−1.23
Emotional manipulation	0.15*	0.46[‡]	−3.93[‡]
Commitment manipulation	0.04	0.23[‡]	−2.20*
Derogation of competitors	0.14*	0.18[†]	−0.47
Resource display	−0.05	0.12*	−1.94*
Sexual inducements	0.16[†]	0.33[‡]	−2.06*
Appearance enhancement	−0.01	0.12*	−1.48
Love and care	−0.10	−0.05	−0.57
Submission and debasement	0.09	0.24[‡]	−1.75*
Verbal possession signals	0.07	0.07	0.00
Physical possession signals	−0.04	0.06	−1.14
Possessive ornamentation	0.03	0.22[‡]	−2.20*
Derogation of mate	0.11	0.05	0.69
Intrasexual threats	0.17[†]	0.30[‡]	−1.56
Violence against rivals	0.14*	0.04	1.14

Note: $N = 246$ for study 1, $N = 276$ for study 2. Difference between correlations was assessed via Fisher's *r*-to-*z* transformation procedure.

*$P < 0.05$; [†] $P < 0.01$; [‡] $P < 0.001$ (two-tailed).

are known to underreport sexual coercion of intimate partners (e.g. Dobash *et al.*, 1998), and because we measured whether it occurred in the current relationship and not whether it had ever occurred in any relationship.

Appendix 5.2   Study 2: women's partner-reports

Men's self-reports of their violence, controlling behavior, and sexual coercion may not be accurate depictions of reality (Dobash *et al.*, 1998; Edleson & Brygger, 1986; Magdol *et al.*, 1997; O'Leary & Arias, 1988; Shackelford *et al.*, 2005b). The reliability of men's reports of their sexual coercion and mate-retention behaviors, in particular, can be questioned on several fronts. First, men may be reluctant to report their sexual coercion, or if they do, they may underreport the most severe forms of sexual coercion (e.g. Dobash *et al.*, 1998; Edleson & Brygger, 1986). Second, men sometimes underreport their mate-retention behaviors and controlling behaviors, whereas women report these behaviors with relative accuracy (e.g. Magdol *et al.*, 1997; Shackelford *et al.*, 2005b). Women's reports of their partners' sexual coercion and mate-retention behaviors may reflect more accurately the incidence of such behaviors. In addition, men's perceptions of their partner's infidelities may not be accurate. Using an independent sample of women in committed, sexual relationships, study 2 examined women's observations of their partner's sexual coercion and mate-retention behaviors. These independent reports offered an additional test of the hypothesis tested in study 1.

METHODS

Participants

Two hundred and seventy six women, each of whom was in a committed, sexual relationship with a man for at least 1 year, participated in this study. About half of the participants were drawn from a university in southern Florida and the other half from surrounding communities. The mean age of the participants was 22.3 years (SD = 5.7 years), the mean age of the participants' partners was 24.4 years (SD = 6.9 years), and the mean relationship length was 41.3 months (SD = 39.6 months). None of the women in study 2 were partners of the men who participated in study 1, making the two studies independent.

Materials

The survey for study 2 paralleled the one used in study 1. Participants in study 2 reported on their past infidelities and likelihood of committing future infidelities, their partner's mate-retention behaviors, and their partners' sexual coercion in the current relationship.

Procedure

Paralleling study 1, three criteria must have been met to qualify for participation. The prospective participant must be (1) female, (2) at least 18 years of age, and (3) currently involved in a committed, sexual relationship with a man for at least 1 year. Upon the prospective participant's arrival at the scheduled time and location, the researcher confirmed that the prospective participant met the three participation criteria. The same procedure was followed as in study 1.

RESULTS AND DISCUSSION

As in study 1, we created the composite variable *female infidelity* ($\alpha = 0.74$) by summing four variables: (1) past sexual infidelity, (2) past emotional infidelity, (3) likelihood of future sexual infidelity, and (4) likelihood of future emotional infidelity.

We then calculated scores for men's sexual coercion using women's responses to the SCIRS (Shackelford & Goetz, 2004). The alpha reliability for the summed responses to the 34 items of the SCIRS was $\alpha = 0.97$. We calculated scores for 19 mate-retention tactics using the 104 items in the MRI (Buss, 1988; Shackelford et al., 2005a). The alpha reliability for the entire inventory (with responses summed across items) was $\alpha = 0.97$. Alpha reliabilities for the 19 mate-retention tactics varied in this study from $\alpha = 0.50$ to 0.91, with a mean of $\alpha = 0.78$.

Hypothesis 1 stated that men's sexual coercion in the context of an intimate relationship is related positively to their partner's infidelities. Consistent with this hypothesis, women's reports of their partner's sexual coercion correlated positively with their infidelities; $r(274) = 0.32$ ($P < 0.001$). Women who reported that they had been or were likely to be unfaithful reported that their partners used more sexual coercion.

Hypothesis 2 stated that men's sexual coercion in the context of an intimate relationship is related positively to their mate-retention behaviors. Consistent with this hypothesis, women's reports of their partners' sexual coercion correlated positively with their partners use of mate-retention behaviors; $r(274) = 0.34$ ($P < 0.001$). Women who reported that their partner used more sexual coercion in their relationship reported that their partners used more mate-retention behaviors. Although the focus of hypothesis 2 was the relationship between men's sexual coercion and their broad use of mate-retention behaviors, we also report in Table 5.1 correlations between women's reports of their partners' sexual coercion and the 19 mate-retention tactics identified by Buss (1988) and his colleagues (Buss & Shackelford, 1997; Shackelford et al.,

2005a). Fourteen of the 19 mate-retention tactics showed significant positive correlations with women's reports of their partners' use of sexual coercion in the current relationship: vigilance, concealment of mate, monopolization of time, jealousy induction, punish mate's infidelity threat, emotional manipulation, commitment manipulation, derogation of competitors, resource display, sexual inducements, appearance enhancement, submission and debasement, possessive ornamentation, and intrasexual threats.

As in study 1 a secondary goal of study 2 was to obtain a prevalence estimate of FIPC in a sample of young adults. We asked women if their current partners had ever physically forced them to have sex or if their current partners had ever initiated sex with them when they were unaware (e.g. when they were asleep, drunk, or on medication) and continued against their will. Twenty five of 276 of women in this sample (9.1%) admitted that they had experienced at least one FIPC by their current partner. The prevalence of FIPC in this study approaches Finkelhor and Yllo's (1985) figure of 10% despite the fact that we measured whether it occurred in the current relationship and not whether it has ever occurred in any relationship.

References

Alexander, R. and Noonan, K. (1979). Concealment of ovulation, parental care, and human social evolution. In N. Chagnon and W. Irons, eds., *Evolutionary Biology and Human Social Behavior*. North Scituate, MA: Duxbury Press, pp. 436–53.

Andersson, M. (1994). *Sexual Selection*. Princeton, NJ: Princeton University Press.

Bailey, R. O., Seymour, N. R., and Stewart, G. R. (1978). Rape behavior in blue-winged teal. *Auk*, **95**, 188–90.

Baker, R. R. and Bellis, M. A. (1993). Human sperm competition: ejaculate adjustment by males and the function of masturbation. *Animal Behaviour*, **46**, 861–85.

Barash, D. P. (1977). Sociobiology of rape in mallards (*Anas platyrhynchos*): response of the mated male. *Science*, **197**, 788–9.

Bart, P. (1975). Rape doesn't end with a kiss. *Viva*, 40–42, 101–7.

Bergen, R. K. (1996). *Wife Rape: Understanding the Response of Survivors and Service Providers*. Thousand Oaks, CA: Sage.

Birkhead, T. R., Hunter, F. M., and Pellatt, J. E. (1989). Sperm competition in the zebra finch, *Taeniopygia guttata*. *Animal Behaviour*, **38**, 935–50.

Bowker, L. H. (1983). Marital rape: a distinct syndrome? *Social Casework: the Journal of Contemporary Social Work*, **64**, 347–52.

Buss, D. M. (1988). From vigilance to violence: tactics of mate retention in American undergraduates. *Ethology and Sociobiology*, **9**, 291–317.

Buss, D. M. and Shackelford, T. K. (1997). From vigilance to violence: mate retention tactics in married couples. *Journal of Personality and Social Psychology*, **72**, 346–61.

Campbell, J. C. (1989). Women's responses to sexual abuse in intimate relationships. *Women's Health Care International*, **8**, 335–47.

Campbell, J. C. and Soeken, K. L. (1999). Forced sex and intimate partner violence: effects on women's risk and women's health. *Violence Against Women*, **5**, 1017–35.

Cheng, K. M., Burns, J. T., and McKinney, F. (1983). Forced copulation in captive mallards: III. Sperm competition. *The Auk*, **100**, 302–10.

Dobash, R. E., Dobash, R. P., Cavanagh, K., and Lewis, R. (1998). Separate and intersecting realities: a comparison of men's and women's accounts of violence against women. *Violence Against Women*, **4**, 382–414.

Edleson, J. and Brygger, M. (1986). Gender differences in reporting of battering incidents. *Family Relations*, **35**, 377–82.

Finkelhor, D. and Yllo, K. (1985). *License to Rape: Sexual Abuse of Wives*. New York: Holt, Rinehart, & Winston.

Frieze, I. H. (1983). Investigating the causes and consequences of marital rape. *Signs: Journal of Women in Culture and Society*, **8**, 532–53.

Gallup, G. G., Burch, R. L., Zappieri, M. L., Parvez, R. A., Stockwell, M. L. and Davis, J. A. (2003). The human penis as a semen displacement device. *Evolution and Human Behavior*, **24**, 277–89.

Gangestad, S. W., Thornhill, R., and Garver, C. E. (2002). Changes in women's sexual interests and their partner's mate-retention tactics across the menstrual cycle: evidence for shifting conflicts of interest. *Proceedings of the Royal Society of London B*, **269**, 975–82.

Goetz, A. T. and Shackelford, T. K. (2006). Sexual coercion and forced in-pair copulation as sperm competition tactics in humans. *Human Nature* (in press).

Goetz, A. T., Shackelford, T. K., Weekes-Shackelford, V. A., *et al*. (2005). Mate retention, semen displacement, and human sperm competition: a preliminary investigation of tactics to prevent and correct female infidelity. *Personality and Individual Differences*, **38**, 749–63.

Goodwin, D. (1955). Some observations on the reproductive behavior of rooks. *British Birds*, **48**, 97–107.

Hall, G. C. N., Shondrick, D. D., and Hirschman, R. (1993). The role of sexual arousal in sexually aggressive behavior: a meta-analysis. *Journal of Consulting and Clinical Psychology*, **61**, 1091–95.

Koss, M. P. and Oros, C. J. (1982). Sexual experiences survey: a research instrument investigating sexual aggression and victimization. *Journal of Consulting and Clinical Psychology*, **50**, 455–7.

Lalumière, M. L. and Quinsey, V. L. (1994). The discriminability of rapists from non-sex offenders using phallometric measures: a meta-analysis. *Criminal Justice and Behavior*, **21**, 33–48.

Lalumière, M. L., Harris, G. T., Quinsey, V. L., and Rice, M. E. (2005). *The Causes of Rape: Understanding Individual Differences in Male Propensity for Sexual Aggression*. Washington, DC: APA Press.

Magdol, L., Moffitt, T. E., Caspi, A., Newman, D. L., Fagan, J., and Silva, P. A. (1997). Gender differences in partner violence in a birth cohort of 21-year-olds: bridging the gap between clinical and epidemiological approaches. *Journal of Consulting and Clinical Psychology*, **65**, 68–78.

McKinney, F. and Stolen, P. (1982). Extra-pair-bond courtship and forced copulation among captive green-winged teal (*Anas crecca carolinensis*). *Animal Behaviour*, **30**, 461–74.

McKinney, F., Derrickson, S. R., and Mineau, P. (1983). Forced copulation in waterfowl. *Behavior*, **86**, 250–94.

McKinney, F., Cheng, K. M., and Bruggers, D. J. (1984). Sperm competition in apparently monogamous birds. In R. L. Smith, ed., *Sperm Competition and Evolution of Animal Mating Systems*. New York: Academic Press, pp. 523–45.

O'Leary, K. D. and Arias, I. (1988). Assessing agreement of reports of spouse abuse. In G. T. Hotaling, D. Finkelhor, J. T. Kirkpatrick, and M. Straus, eds., *Family Abuse and its Consequences*. Newbury Park, CA: Sage, pp. 218–27.

Pagelow, M. D. (1981). *Woman-Battering: Victims and their Experiences*. Beverly Hills, CA: Sage.

Pagelow, M. D. (1988). Marital rape. In V. B. Van Hasselt, R. L. Morrison, A. A. Bellack, and M. Hersen, eds, *Handbook of Family Violence*. New York: Plenum Press, pp. 207–32.

Painter, K. and Farrington, D. P. (1999). Wife rape in Great Britain. In R. Muraskin, ed., *Women and Justice: Development of International Policy*. New York: Gordon and Breach, pp. 135–64.

Palmer, C. T. (1991). Human rape: adaptation or by-product? *Journal of Sex Research*, **28**, 365–86.

Parker, G. A. (1970). Sperm competition and its evolutionary consequences in the insects. *Biological Reviews*, **45**, 525–67.

Pound, N. (2002). Male interest in visual cues of sperm competition risk. *Evolution and Human Behavior*, **23**, 443–66.

Russell, D. E. H. (1982). *Rape in Marriage*. New York: Macmillan Press.

Seymour, N. R. and Titman, R. D. (1979). Behaviour of unpaired male black ducks (*Anas rupribes*) during the breeding season in a Nova Scotia tidal marsh. *Canadian Journal of Zoology*, **57**, 2412–28.

Shackelford, T. K. and Buss, D. M. (1997). Cues to infidelity. *Personality and Social Psychology Bulletin*, **23**, 1034–45.

Shackelford, T. K. and Goetz, A. T. (2004). Men's sexual coercion in intimate relationships: development and initial validation of the Sexual Coercion in Intimate Relationships Scale. *Violence and Victims*, **19**, 21–36.

Shackelford, T. K. and LeBlanc, G. J. (2001). Sperm competition in insects, birds, and humans: insights from a comparative evolutionary perspective. *Evolution and Cognition*, **7**, 194–202.

Shackelford, T. K., LeBlanc, G. J., Weekes-Shackelford, V. A., Bleske-Rechek, A. L., Euler, H. A., and Hoier, S. (2002). Psychological adaptation to human sperm competition. *Evolution and Human Behavior*, **23**, 123–38.

Shackelford, T. K., Goetz, A. T., LaMunyon, C. W., Quintus, B. J., and Weekes-Shackelford, V. A. (2004). Sex differences in sexual psychology produce sex similar preferences for a short-term mate. *Archives of Sexual Behavior*, **33**, 405–12.

Shackelford, T. K., Goetz, A. T., and Buss, D. M. (2005a). Mate retention in marriage: further evidence of the reliability of the Mate Retention Inventory. *Personality and Individual Differences*, **39**, 415–25.

Shackelford, T. K., Goetz, A. T., Buss, D. M., Euler, H. A., and Hoier, S. (2005b). When we hurt the ones we love: predicting violence against women from men's mate retention. *Personal Relationships*, **12**, 447–63.

Shackelford, T. K., Pound, N., and Goetz, A. T. (2005c). Psychological and physiological adaptation to human sperm competition. *Review of General Psychology*, **9**, 228–48.

Shackelford, T. K., Goetz, A. T., Guta, F. E., and Schmitt, D. P. (2006). Mate guarding and frequent in-pair copulation in humans: complementary anti-cuckoldry tactics. *Human Nature*. (in press).

Shields, N. M. and Hanneke, C. R. (1983). Battered wives' reactions to marital rape. In R. Gelles, G. Hotaling, M. Straus, and D. Finkelhor, eds., *The Dark Side of Families*. Beverly Hills, CA: Sage, pp. 131–48.

Shields, W. M. and Shields, L. M. (1983). Forcible rape: an evolutionary perspective. *Ethology and Sociobiology*, **4**, 115–36.

Smith, R. L. (1984). Human sperm competition. In R. L. Smith, ed., *Sperm Competition and the Evolution of Animal Mating Systems*. New York: Academic Press, pp. 601–60.

Thornhill, R. and Palmer, C. T. (2000). *A Natural History of Rape*. Cambridge, MA: MIT Press.

Thornhill, R. and Thornhill, N. W. (1983). Human rape: an evolutionary analysis. *Ethology and Sociobiology*, **4**, 137–73.

Thornhill, R. and Thornhill, N. W. (1992). The evolutionary psychology of men's coercive sexuality. *Behavioral and Brain Sciences*, **15**, 363–421.

Valera, F., Hoi, H., and Kristin, A. (2003). Male shrikes punish unfaithful females. *Behavioral Ecology*, **14**, 403–8.

Walker, L. E. (1979). *The Battered Woman*. New York: Harper & Row.

Weis, K. and Borges, S. S. (1973). Victimology and rape: the case of the legitimate victim. *Issues in Criminology*, **8**, 71–115.

Wilson, M. and Daly, M. (1992). The man who mistook his wife for a chattel. In J. H. Barkow, L. Cosmides, and J. Tooby, eds., *The Adapted Mind*. New York: Oxford University Press, pp. 289–322.

Wyckoff, G. J., Wang, W., and Wu, C. (2000). Rapid evolution of male reproductive genes in the descent of man. *Nature*, **403**, 304–8.

PART III   INTRAVAGINAL TACTICS: SPERM
COMPETITION AND SEMEN
DISPLACEMENT

6

Sperm competition and its evolutionary consequences in humans

AARON T. GOETZ AND TODD K. SHACKELFORD
Florida Atlantic University

Identifying sperm competition

Sexual selection is the mechanism that favors an increase in the frequency of alleles associated with reproduction (Darwin, 1871). Darwin distinguished sexual selection from natural selection, but today most evolutionary scientists combine the two concepts under the name, natural selection. Sexual selection is composed of intrasexual competition (competition between members of the same sex for sexual access to members of the opposite sex) and intersexual selection (differential mate choice of members of the opposite sex). Focusing mainly on precopulatory adaptations associated with intrasexual competition and intersexual selection, postcopulatory sexual selection was largely ignored even a century after the presentation of sexual selection theory. Parker (1970) was the first to recognize that male–male competition may continue even after the initiation of copulation when males compete for fertilizations. More recently, Thornhill (1983) and others (e.g. Eberhard, 1996) recognized that intersexual selection may also continue after the initiation of copulation when a female biases paternity between two or more males' sperm. The competition between males for fertilization of a single female's ova is known as *sperm competition* (Parker, 1970), and the selection of sperm from two or more males by a single female is known as *cryptic female choice* (Eberhard, 1996; Thornhill, 1983). Although sperm competition and cryptic female choice together compose postcopulatory sexual selection (see Table 6.1), sperm competition is often used in reference to both processes (e.g. Baker & Bellis, 1995; Birkhead & Møller, 1998; Simmons, 2001; Shackelford, Pound, & Goetz, 2005). In this chapter, we review the current state of knowledge regarding human sperm competition (and see Shackelford *et al.*, 2005).

Table 6.1. *Precopulatory and postcopulatory sexual selection.*

	Sexual selection	
	Intrasexual competition	Intersexual selection
Precopulatory	How do males compete for mates?	How do females select mates?
Postcopulatory	If two or more males have copulated with a female, how do males compete for fertilizations? (sperm competition)	If a female has copulated with two or more males, how does she select sperm? (cryptic female choice)

Sperm competition in non-human species

Examining the adaptive problems non-human species faced and the resultant evolved solutions to these problems can often provide insight into the adaptive problems and evolved solutions in humans (and vice versa). Shackelford and LeBlanc (2001) argued that because humans share similar adaptive problems with insects (e.g. mate retention) and birds (e.g. extra-pair copulations), humans, insects, and birds may share similar solutions to these adaptive problems. Shackelford and LeBlanc (2001) argued that applying a comparative evolutionary psychological approach to the study of evolved solutions to problems of sperm competition may lead to a better understanding of human sperm competition. We will therefore review some of the first work on sperm competition relevant to humans.

In species with internal fertilization, there exists the potential for sperm competition whenever a female mates with multiple males in a sufficiently short period of time so that live sperm from two or more males are present in her reproductive tract. One of the first hypotheses generated by sperm-competition theory was that males will deliver more sperm when the risk of sperm competition is high (Parker, 1982, 1990a). Across species, therefore, investment in sperm production is predicted to depend on the risk of sperm competition. Within species, males are predicted to allocate their sperm in a prudent fashion and inseminate more sperm when the risk of sperm competition is higher. In accordance with hypotheses generated by sperm-competition theory, investment in sperm production is greater in species for which the risk of sperm competition is higher. In primates (Harcourt *et al.*, 1981; Harvey & Harcourt, 1984; Short, 1979), birds (Møller, 1988a), ungulates (Ginsberg & Rubenstein, 1990), frogs (Jennions & Passmore, 1993), and butterflies (Gage, 1994) testis size (an index of investment in sperm production) is correlated positively with the frequency with which females engage in polyandrous matings. Recent work, in addition, has demonstrated experimentally that exposure to mating environments with high levels of sperm competition can produce significant increases in testis size after only 10 generations in yellow dung flies (*Scathophaga stercoraria*; Hosken & Ward, 2001).

In addition to the evidence that investment in sperm production depends on the risk of sperm competition across species, evidence is accumulating that individual males are capable of prudent sperm allocation (for reviews see Parker *et al.*, 1997; Wedell, Gage, & Parker, 2002). Experiments have demonstrated that males in many species are capable of adjusting the number of sperm they deliver from one insemination to the next in response to cues of sperm-competition risk or intensity. Males need to rely on cues predictive of sperm-competition risk because this risk often cannot be assessed directly. Accordingly, there is experimental evidence that males of various species respond to cues of elevated sperm-competition risk in an adaptive fashion. Some of the cues used include male mating status, in species where it predicts the likelihood of mating with an already-mated female (Cook & Wedell, 1996), and female mating status where it is detectable (Gage & Barnard, 1996). In addition, males of various species appear to be sensitive to the mere presence of one or more rival males during a particular mating event. Field crickets (*Gryllodes supplicans*) and house crickets (*Acheta domesticus*), for example, increase the number of sperm they inseminate in proportion to the number of rivals present (Gage & Barnard, 1996). Of perhaps most relevance to the work on the responses of human males to cues of sperm competition risk is the finding that male rats (*Rattus norvegicus*) adjust the number of sperm they inseminate depending on the amount of time they have spent with a particular female prior to copulation (Bellis, Baker, & Gage, 1990). In addition, male rats inseminate more sperm when mating in the presence of a rival male (Pound & Gage, 2004).

For species that practice social monogamy, the mating system in which males and females form long-term pair bonds but also pursue extra-pair copulations, extra-pair copulations by females create the primary context for sperm competition (Birkhead & Møller, 1992; Smith, 1984). A male whose female partner engages in an extra-pair copulation is at risk of cuckoldry and its associated reproductive costs. These reproductive costs include loss of the time, effort, and resources the male spent attracting his partner, the potential misdirection of his current and future resources to a rival's offspring, and the loss of his mate's investment in any offspring he may have had with her in the future (Buss, 2004; Trivers, 1972). Because cuckoldry is so reproductively costly, males of paternally investing species are expected to possess adaptations that decrease the likelihood of being cuckolded.

Anti-cuckoldry tactics fall into three categories: *preventative tactics*, designed to minimize female infidelity; *sperm-competition tactics*, designed to minimize conception in the event of female infidelity; and *differential paternal investment*, designed to allocate paternal investment prudently in the event that female infidelity may have resulted in conception (Platek, 2003; Shackelford *et al.*, 2000; Wilson & Daly, 1992). It is expected that for a given mating interaction, the performance of these tactics will tend to proceed in the sequence outlined above. A male's best strategy is to prevent female infidelity and, if he is unsuccessful in preventing female infidelity, he would benefit by attempting to prevent conception by a rival male. If he is unsuccessful in preventing conception by a rival male, he would benefit by adjusting paternal effort according to available paternity cues. The performance of one tactic does not necessitate the neglect of another tactic; indeed, a reproductively wise strategy would be to perform all three categories of anti-cuckoldry tactics.

Male swallows (*Hirundo rustica*), a socially monogamous species, have been observed performing preventative tactics, sperm-competition tactics, and differential paternal investment (Møller, 1985, 1987, 1988b; cited in Wilson & Daly, 1992). Male swallows guard their mates while they are fertile (Møller, 1987); they adjust their rate of in-pair copulation relative to the proximity of rival males (Møller, 1985); and they adjust paternal effort according to the observed frequency of their mate's extra-pair copulations (Møller, 1988b). Employing preventative tactics, sperm-competition tactics, and differential investment to avoid the costs associated with cuckoldry, male swallows, as well as males of other bird species, possess an arsenal of anti-cuckoldry tactics. Before we examine men's anti-cuckoldry tactics and other adaptive solutions created by sperm competition, we must first consider whether sperm competition was an important selection pressure for humans.

Has sperm competition been an important selection pressure for humans?

The likelihood and selective importance of sperm competition in humans are issues of scholarly debate and controversy. Smith (1984) argued that the comparatively large size of the human penis, and the fact that human testes are somewhat larger in relation to body size than are those of monogamous primates (Short, 1981), suggests that sperm competition has been a recurrent feature of human evolutionary history. Smith (1984) argued that facultative polyandry (i.e. female sexual infidelity) would have been the most common reason for the simultaneous presence of live sperm from two or more men in the reproductive tract of an ancestral woman. Smith (1984) acknowledged that other contexts in which sperm competition might have occurred include consensual communal sex, courtship, rape, and prostitution, but argued that these contexts may not have occurred with sufficient frequency over human evolutionary history to provide selection pressures for adaptations to sperm competition equivalent to female infidelity.

Evidence of an evolutionary history of female infidelity and sperm competition is provided by the ubiquity and power of male sexual jealousy. Male sexual jealousy could only evolve if female sexual infidelity was a recurrent feature of human evolutionary history (see e.g. Buss et al., 1992; Daly, Wilson, & Weghorst, 1982; Symons, 1979), and female infidelity increases the likelihood that sperm from two or more men recurrently occupied the reproductive tract of a single woman. Indeed, based on past and present infidelity rates of men and women, it may be concluded that humans practice social monogamy. Because of female sexual infidelity, members of socially monogamous species are likely to face the adaptive problems associated with sperm competition (Birkhead & Møller, 1992; Smith, 1984).

Recent genetic studies provide additional evidence for a long evolutionary history of sperm competition in humans. Investigating genes that code for proteins involved in the production and function of sperm, Wyckoff, Wang, and Wu (2000) found that these genes have been evolving at a much faster rate than most other human genes. Wyckoff et al. (2000) concluded that this rapid genetic change could only have occurred if ancestral women had concurrent sexual partners often enough that sperm of different men competed to fertilize a woman's eggs.

Those questioning the application of sperm competition to humans (e.g. Birkhead, 2000; Dixson, 1998; Gomendio, Harcourt, & Roldán, 1998) contend that sperm competition in humans, although possible, may not be as intense as in other species with adaptations to sperm competition. Recent work on the psychological, behavioral, and anatomical evidence of human sperm competition (reviewed in this chapter), however, was not considered in these previous critiques of human sperm competition. When considering all of the evidence of adaptations to sperm competition in men and in women, it is reasonable to conclude that sperm competition is likely to have been a recurrent and selectively important feature of human evolutionary history.

DO WOMEN GENERATE SPERM COMPETITION?

Evolutionary accounts of human sexual psychology have tended to emphasize the benefits to men of short-term mating and sexual promiscuity (e.g. Buss & Schmitt, 1993; Symons, 1979). For men to pursue short-term sexual strategies, however, there must be women who mate non-monogamously

(Greiling & Buss, 2000). Moreover, if ancestral women never engaged in short-term mating, men could not have evolved a strong desire for sexual variety (Schmitt, Shackelford, & Buss, 2001; Schmitt *et al.*, 2001, 2003; Smith, 1984).

Ancestral women may have benefited from facultative polyandry in several ways (Smith, 1984; for a review, see Greiling & Buss, 2000). Some of the most important potential benefits include the acquisition of resources, either in direct exchange for sex with multiple men (Symons, 1979) or by creating paternity confusion as a means to elicit investment (Hrdy, 1981). Alternatively, ancestral women may have benefited indirectly by accepting resources and parental effort from a primary mate while copulating opportunistically with men of superior genetic quality (Smith, 1984; Symons, 1979). Furthermore, extra-pair sex might have been useful as insurance against the possibility that a primary mate was infertile, and in unpredictable environments it may be advantageous for women to ensure that offspring are sired by different men and are thus genetically diverse (Smith, 1984). Jennions and Petrie (2000) provide a comprehensive review of the genetic benefits to females of multiple mating.

Multiple mating by women is a prerequisite for sperm competition to occur, but not all patterns of polyandry will generate postcopulatory competition between men. For sperm competition to occur, women must copulate with two or more men in a sufficiently short period of time such that there is overlap in the competitive life spans of the rival ejaculates. The length of this competitive "window" might be as short as 2–3 days (Gomendio & Roldán 1993), or as long as 7–9 days (Smith, 1984). Using an intermediate estimate of 5 days, Baker and Bellis (1995) argued that the questionnaire data they collected on female sexual behavior indicated that 17.5% of British women "double-mated" in such a way as to generate sperm competition (in the absence of barrier contraception) at some point during the first 50 copulations in their lifetimes. Although questions have been posed about the accuracy of this estimate (e.g. Gomendio *et al.*, 1998), it is clear that women in contemporary human populations do frequently mate in a polyandrous fashion and thus potentially generate sperm competition in their reproductive tracts.

Large-scale studies of sexual behavior have not collected data on the frequency with which women double-mate specifically, but many have recorded how often they engage in concurrent sexual relationships more generally. Laumann *et al.* (1994), for example, found that 83% of respondents who report having had five or more sexual partners in the past year also report that at least two of these relationships were concurrent. Not all concurrent sexual relationships involve copulations with different men within a sufficiently short space of time to be considered double-matings, but it is likely that many do. For this reason, some researchers have argued that the rate at which women participate in concurrent sexual relationships provides an index of the likelihood of sperm competition in a population. Gomendio *et al.* (1998), for example, argued that survey data indicate that only 2% of women in Britain have engaged in concurrent sexual relationships in the past year and, consequently, that sperm competition is likely to be a relatively infrequent occurrence. However, a major study of sexual behavior in Britain – the National Survey of Sexual Attitudes and Lifestyles conducted between 1999 and 2001 (Johnson *et al.*, 2001) – revealed that 9% of women overall, and 15% of those aged 16–24 years, reported having had concurrent sexual relationships with men during the preceding year.

It is likely that women's sexual behavior does sometimes generate sperm competition. Bellis and Baker (1990) argued that women "schedule" or time their copulations in a way that *actively promotes* sperm competition. Active promotion of successive insemination by two or more men may allow a woman to be fertilized by the most competitive sperm. Bellis and Baker (1990) documented that women are more likely to double-mate when the probability of conception is highest, suggesting that women may promote sperm competition. When the probability of conception is lower, in contrast, women separate in time in-pair and extra-pair copulations over a 5-day period, making sperm competition less likely. Bellis and Baker (1990) argued that the results cannot be attributed to men's preferences for copulation with women at peak fertility. According to Bellis and Baker (1990), if the results were due to men's preferences for copulation during peak fertility and not to women's active promotion of sperm competition, then in-pair copulations should occur more often during fertile phases of the menstrual cycle, just as was found for extra-pair copulations.

Bellis and Baker (1990) may have been too quick to dismiss the possibility that men prefer to copulate with a woman during peak fertility, however. Because women may be attempting to secure genetic benefits from their extra-pair partners (see Gangestad & Simpson, 2000; Greiling & Buss, 2000), women are predicted to prefer to copulate with extra-pair partners when conception is highest. A woman might simultaneously avoid copulation with an in-pair partner while seeking extra-pair sex. So, although her in-pair partner might prefer to copulate with her precisely during the peak fertility phase of her cycle, this may not be reflected in her actual pattern of copulations. Therefore, Bellis and Baker's (1990) finding that women are more likely to double-mate when the probability of conception is highest is consistent with the hypothesis that women sometimes actively promote sperm competition, but does not rule out the possibility that both in-pair and extra-pair partners prefer to copulate with a woman during her peak fertility.

POLYANDROUS SEX IN WOMEN'S FANTASIES

Sexual fantasy may provide a "window" through which to view the evolved psychological mechanisms that motivate sexual behavior (Ellis & Symons, 1990; Symons, 1979). A large empirical literature has addressed sex differences in sexual fantasy, and much of this work has been conducted from an evolutionarily informed perspective (see, e.g. Ellis & Symons, 1990; Wilson, 1987, 1997; Wilson & Lang, 1981; and see Leitenberg & Henning, 1995, for a broad review of empirical work on sexual fantasy). This work documents several marked sex differences in the content of sexual fantasies, consistent with hypotheses generated from Trivers' (1972) theory of parental investment and sexual selection. For example, given the asymmetric costs associated with sexual reproduction, sexual access to mates limits reproductive success for males more than for females. Consequently, it has been hypothesized that men more than women will have sexual fantasies that involve multiple, anonymous sexual partners who do not require an investment of time, energy, or resources prior to granting sexual access (e.g. Ellis & Symons, 1990), and empirical investigations have confirmed this hypothesis. Indeed, one of the largest sex differences occurs for fantasies about having sex with two or more members of the opposite sex concurrently, with men more than women reporting this fantasy (see review in Leitenberg & Henning, 1995).

Tests of the hypothesis that men more than women fantasize about concurrent sex with two or more partners have inadvertently provided data on women's polyandrous sexual fantasies. Although this work clearly indicates that men are more likely than women to report fantasies of concurrent sex with multiple partners, polyandrous sex is a recurring theme for some women. In a large survey study, for example, Hunt (1974) found that 18% of women report fantasies of polyandrous sex, imagining themselves as a woman having sex with two or more men concurrently. Wilson (1987) surveyed nearly 5000 readers of Britain's top-selling daily newspaper about their favorite sexual fantasy and performed content analyses on the responses of a random subsample of 600 participants. Polyandrous sex was the key element of the favorite sexual fantasy reported by 15% of female participants.

Studies using smaller samples of participants also provide evidence that polyandry is a common theme of women's sexual fantasies, albeit less common than for men. For example, Rokach (1990) reported that, although sex with more than one partner accounted for 14% of the sexual fantasies reported by a sample of 44 men, it accounted for 10% of the fantasies reported by a sample of 54 women. Person et al. (1989) and Pelletier and Herold (1988) documented that 27 and 29%, respectively, of the women sampled report fantasies of polyandrous sex. And fully 41% of women sampled by Arndt, Foehl, and Good (1985) report fantasies involving sex with two men at the same time, and Price and Miller (1984) report that polyandrous sex was among the 10 most frequently reported fantasies in a small sample of college women. Indeed, polyandrous sex ranked as the third most frequent fantasy of black women and as the eighth most frequent fantasy of white women in this study.

If sexual fantasy reflects sexual desires and preferences that might sometimes be acted upon, then previous research indicates that polyandrous sex is not an unlikely occurrence, particularly given the well-established finding that women more than men are the "gatekeepers" of sexual access – including when, where, and the conditions under which sex occurs (see, e.g. Buss, 2004; Symons, 1979). If, as Symons (1979) has argued (and see Buss, 2004; Ellis & Symons, 1990), sexual fantasy provides a window through which to view evolved human psychology, then human female sexual psychology may include design features dedicated to the pursuit of polyandrous sex, with the consequence of promoting sperm competition.

Men's adaptations to sperm competition

Sperm competition can take one of two forms: *contest competition*, in which rival ejaculates actively interfere with each other's ability to fertilize an ovum or ova, and *scramble competition*, which is more akin to a simple race or lottery. In mammals, there are theoretical reasons to believe that most sperm competition takes the form of a scramble, and modeling studies and experimental findings support this view (Gomendio *et al.*, 1998). Male adaptations to scramble competition are likely to take the form of physiological, anatomical, and behavioral features that increase the male's chances of fertilizing an ovum or ova in a competitive environment in which the ability to deliver large numbers of sperm is a crucial determinant of fertilization success.

IS THERE EVIDENCE OF PRUDENT SPERM ALLOCATION BY MEN?

Sperm-competition theory can be used to generate the predictions that, across species, investment in sperm production will depend on the level of sperm competition, and that, where the risk of sperm competition is *variable*, individual males will allocate their sperm in a prudent fashion and will, accordingly, inseminate more sperm when the risk is higher (Parker, 1982, 1990a, 1990b). It is possible that adaptations to *variable* levels of sperm competition will be seen in species where overall levels are not especially high – but where sperm competition is a sufficiently frequent occurrence to select for mechanisms that allow prudent sperm allocation.

Compared to other primates, human ejaculates do not contain especially large numbers of sperm (Baker & Bellis, 1995; Dixson, 1998). Men, therefore, do not appear be adapted to particularly high levels of sperm-competition. Nevertheless, it may be the case that men have physiological adaptations that allow them to allocate sperm prudently in the face of variable levels of sperm-competition risk. The only published evidence, however, indicating that men can adjust ejaculate composition in response to adaptively relevant aspects of their environment was provided by Baker and Bellis (1989a, 1993).

Baker and Bellis first reported that the number of sperm inseminated by men varied according to hypotheses generated by sperm-competition theory (Baker & Bellis, 1989a). For this study, 10 heterosexual couples provided semen specimens collected via masturbation and others collected during copulation. Although participants provided multiple specimens, the analysis was restricted to the first specimen provided in each of the two experimental contexts (masturbatory and copulatory). For the 10 copulatory specimens, there was a significant negative rank-order correlation between the percentage of time the couple had spent together since their last copulation and the estimated number of sperm in the ejaculate. That is, men who had spent the most time apart from their partners since their last copulation produced ejaculates containing the most sperm. Because the percentage of time spent apart from a partner is a reliable cue of the risk of female double-mating, these findings are consistent with the hypothesis that there is a positive association between the number of sperm inseminated and the risk of sperm competition (Parker 1970, 1982). What Baker and Bellis (1989a) reported, however, was a between-subjects relationship between sperm-competition risk and ejaculate composition. Baker and Bellis (1989a) did not provide direct evidence of prudent sperm allocation by men from one specimen to the next in response to variation in sperm competition risk. It could be the case that men who tended to produce larger ejaculates also tended to spend a greater proportion of their time between copulations apart from their partners. Moreover, this relationship could be mediated by between-male differences in testicular size and associated levels of testosterone production if variability in these variables predicts semen parameters and certain aspects of sexual behavior.

In a follow-up to this initial report, Baker and Bellis (1993) attempted to address the aforementioned problems by including in their analyses more than one ejaculate from each couple that participated in this second study. Twenty-four couples provided a total of 84 copulatory ejaculates. To assess whether the number of sperm inseminated by a man depended on the percentage of time spent together since the last copulation with his partner, only those copulatory specimens that were preceded by an ejaculation also produced during an in-pair copulation (IPC) were included in the analyses (IPC–IPC ejaculates). Forty specimens produced by five men were included in the final analysis, and for these a non-parametric test based on ranks indicated a negative association between the number of sperm inseminated and the proportion of time the couple had spent together since their last copulation – evidence of prudent sperm allocation by men.

Although data were presented for the first IPC–IPC ejaculates produced by all 15 couples who provided copulatory specimens, an analysis similar to that presented in the 1989 paper was not reported. Shackelford and his colleagues (2005) conducted this analysis using the 1993 data, which revealed that, for the first IPC–IPC ejaculate produced by each couple, the negative rank-order correlation between the number of sperm inseminated by a man and the percentage of time spent together with his partner since their last copulation was marginally statistically significant ($r = -0.50$; $P = 0.058$).

Aside from the small sample size used in Baker and Bellis' (1993) demonstration of prudent sperm allocation by individual men, a number of additional methodological concerns have led some researchers to be skeptical of the findings. One concern is the possibility that the people who participated in this intrusive research about some of their most private behaviors may not be representative of most people. Recruited from the staff and postgraduate students in a biology department, the participants might have had some knowledge of the experimental hypothesis. It is not clear, however, how such knowledge could affect semen parameters. Knowledge about the experimental hypothesis could have affected the sexual behavior of the participants, and there is some evidence that semen parameters are subject to behavioral influences (Pound et al., 2002; Zavos, 1985, 1988; Zavos & Goodpasture 1989; Zavos et al., 1994). However, evidence that men are able to adjust their semen parameters in response to the demand characteristics of an experiment would perhaps be more remarkable than evidence of prudent sperm allocation in the face of cues of sperm-competition risk.

PSYCHOLOGICAL MECHANISMS ASSOCIATED WITH PRUDENT SPERM ALLOCATION

The findings of Baker and Bellis (1989a, 1993) suggest that men may be capable of such prudent sperm allocation, but it is not clear how men accomplish this. Little attention has been paid, however, to the psychological mechanisms that might be involved in regulating such responses. Adaptive changes in semen parameters can serve no function unless they are accompanied by a desire to copulate with a partner when cues of sperm-competition risk are present. Accordingly, Shackelford *et al.* (2002) investigated the psychological responses of men to cues of sperm-competition risk, arguing that there must be psychological mechanisms in men that evolved to motivate behavior that would have increased the probability of success in sperm-competition in ancestral environments.

Baker and Bellis (1993, 1995) operationalized risk of sperm competition as the proportion of time a couple has spent together since their last copulation. The proportion of time spent *apart* since the couple's last copulation is correlated negatively with the proportion of time that they have spent together and is arguably a more intuitive index of the risk of sperm competition and, therefore, Shackelford *et al.* (2002) used this in their work. Shackelford and his colleagues argued that the proportion of time spent apart is information that is processed by male psychological mechanisms that subsequently motivate a man to inseminate his partner as soon as possible, to combat the increased risk of sperm competition.

Total time since last copulation is not clearly linked to the risk of sperm competition. Instead, it is the proportion of time a couple has spent apart since their last copulation – time during which a man cannot account for his partner's activities – that is linked to the risk that his partner's reproductive tract might contain the sperm of rival males (Baker & Bellis, 1995). Nevertheless, total time since last copulation might have important effects on a man's sexual behavior. As the total time since last copulation increases, a man might feel increasingly "sexually frustrated" whether or not that time has been spent apart or together. To address the potential confound, Shackelford *et al.* (2002) assessed the relationships between male sexual psychology and behaviors predicted to be linked to the risk of sperm competition (as assessed by the proportion of time spent apart since last copulation), controlling for the total time since a couple's last copulation.

Shackelford *et al.* (2002) suggested that men might respond differently to cues of sperm-competition risk depending on the nature of their relationship with a particular woman. Satisfaction with, and investment in, a relationship are likely to be linked, with the result that a man who is more satisfied may have more to lose in the event of cuckoldry. For this reason, when examining the responses of men to increases in the proportion of time spent apart from their partner since their last copulation, Shackelford *et al.* controlled for the extent to which the participants were satisfied with their relationships.

Consistent with their predictions, Shackelford *et al.* (2002) found that a man who spends a greater (relative to a man who spends a lesser) proportion of time apart from his partner since the couple's last copulation (and, therefore, faces a higher risk of sperm competition) rates his partner as more attractive, reports that other men find his partner more attractive, reports greater interest in copulating with his partner, and reports that his partner is more interested in copulating with him. Shackelford and his colleagues argued that no existing theory other than sperm-competition theory can account for the predictive utility of the proportion of time spent apart since the couple's last copulation, independent of the total time since last copulation and independent of relationship satisfaction. Additionally, they argued that their findings support the hypothesis that men, like males of other socially monogamous but not sexually exclusive species, have psychological mechanisms designed to solve the adaptive problems associated with a partner's sexual infidelity.

Additionally, psychological mechanisms associated with prudent sperm allocation may explain why men are continually interested in copulating with their partners throughout the duration of a mateship (Klusmann, 2002), a prediction first made by Baker and Bellis (1993). According to Baker and Bellis' (1993) "topping-up" model, a woman's primary partner should desire to maintain an optimum level of sperm in his partner's reproductive tract as a sperm-competition tactic. Surveying German participants, Klusmann (2002) documented that sexual desire for one's partner declines in women but remains constant in men for the duration of a mateship, and interpreted the results in accordance with the topping-up model. Although men report that their sexual satisfaction (Klusmann, 2002) and the quality of marital sex (Chien, 2003) decline with the duration of the mateship, men's desire for sex with their partner does not decline with the duration of the mateship (Klusmann, 2002).

The crux of the topping-up model is that continued sexual desire functions to motivate sexual activity throughout the mateship (i.e. sexual desire without sexual behavior would be an incomplete strategy). Klusmann (2002) found, however, that sexual activity declined in men and women with the duration of the mateship. This finding is not fatal to Klusmann's interpretation of the data or to Baker and Bellis's (1993) model when considering the fact that sexual activity typically requires a consenting partner. Over the duration of a mateship, women (but not men) experience decreased sexual desire and, accordingly, women (but not men) desire sex with their partner less often (Klusmann, 2002). Because women more than men control sexual access, women's waning interest in sex translates into a decrease in sexual activity for both partners.

MEN'S REPRODUCTIVE ANATOMY AND COPULATORY BEHAVIOR

In primates, testis size relative to body weight also is correlated positively with the incidence of polyandrous mating (Harcourt *et al.*, 1981; Harvey &

Harcourt, 1984; Short, 1979). Smith (1984) argued that the fact that men have testes that are larger relative to body size than those of monandrous species such as the gorilla and orangutan suggests that polyandry was an important selection pressure during human evolution. As Gomendio et al. (1998) noted, however, human relative testis size is closer to these monandrous primates than to the highly polyandrous chimpanzee. Nevertheless, Gomendio et al.'s (1998) conclusion that humans are monandrous is not justified. Dichotomizing species into monandrous and polyandrous groups is not useful when there is continuous variation across species in the frequency with which females mate multiply. When the degree of polyandry is considered along a continuum, it is likely that, although human males have not experienced levels of sperm competition as high as have been documented in several primate species, is it unlikely that sperm competition was completely absent over human evolutionary history.

Human males have a penis that is longer than in any other species of ape (Short, 1979), but in relation to body weight it is no longer than the chimpanzee penis (Gomendio et al., 1998). Several arguments have been offered to explain how the length and shape of the human penis might reflect adaptation to an evolutionary history of sperm competition. A long penis may be advantageous in the context of scramble competition, which combines elements of a race and a lottery, because being able to place an ejaculate deep inside the vagina and close to the cervix may increase the chance of fertilization (Baker & Bellis, 1995; Short, 1979; Smith, 1984). Additionally, it has been suggested that the length, width, and shape of the human penis indicate that it may have evolved to function as a semen-displacement device.

Using artificial genitals and simulated semen, Gallup et al. (2003) empirically tested Baker and Bellis's (1995) hypothesis that the human penis may be designed to displace semen deposited by other men in the reproductive tract of a woman. Gallup and his colleagues documented that artificial phalluses that had a glans and a coronal ridge that approximated a real human penis displaced significantly more simulated semen than did a phallus that did not have a glans and a coronal ridge. When the penis is inserted into the vagina, the frenulum of the coronal ridge makes semen displacement possible by allowing semen to flow back under the penis alongside the frenulum and collect on the anterior of the shaft behind the coronal ridge. Displacement of simulated semen only occurred, however, when a phallus was inserted at least 75% of its length into the artificial vagina, suggesting that successfully displacing rival semen may require specific copulatory behaviors. Following allegations of female infidelity or separation from their partners (contexts in which the likelihood of rival semen being present in the reproductive tract is relatively greater), both sexes report that men thrusted deeper and more quickly at the couple's next copulation (Gallup *et al.*, 2003). Such vigorous copulatory behaviors are likely to increase semen displacement.

In an independent test of the hypothesis that successfully displacing rival semen may require specific copulatory behaviors, Goetz *et al.* (2005) investigated whether and how men under a high risk of sperm competition might attempt to "correct" a female partner's sexual infidelity. Using a self-report survey, men in committed sexual relationships reported their use of specific copulatory behaviors arguably designed to displace the semen of rival men. These copulatory behaviors included number of thrusts, deepest thrust, depth of thrusts, on average, and duration of sexual intercourse. An increase in these behaviors would afford a man a better chance to displace rival semen. As hypothesized, men mated to women who place them at a high recurrent risk of sperm competition were more likely to perform semen-displacing behaviors, suggesting that men perform specific copulatory behaviors apparently designed to correct female sexual infidelity by displacing rival semen that may be present in the woman's reproductive tract.

One concern with the hypothesis that the human penis has evolved as a semen-displacement device is that, during copulation, the penis would frequently remove a man's own semen, even if the least conservative estimates of the frequency of extra-pair copulations are accepted. The consequences of such an effect might be minimized, however, if the temporal spacing between successive in-pair copulations is much greater than the spacing between copulations involving different men. Indeed, the refractory period may have been designed for this purpose (Gallup & Burch, 2004). The inability to maintain an erection following ejaculation may function to minimize self-semen displacement.

SPERM COMPETITION AND MEN'S MATE SELECTION

As Baker and Bellis (1995) noted, an evolutionary history of sperm competition may be responsible for myriad male behaviors related directly and indirectly to mating. Research informed by sperm competition theory is just beginning to uncover those behaviors. Aspects of men's short-term mate selection, for example, may have their origins in sperm competition.

To avoid sperm competition or to compete more effectively, men may have evolved mate preferences that function to select as short-term sexual partners women who present the lowest risk of current or future sperm competition (Shackelford *et al.*, 2004). The risk of sperm competition for a man increases with a prospective short-term partner's involvement in one or more relationships. Women who are not in a long-term relationship and do not have casual sexual partners, for example, present a low risk of sperm competition.

Consequently, such women may be perceived as desirable short-term sexual partners. Women who are not in a long-term relationship but who engage in short-term matings may present a moderate risk of sperm competition, because women who engage in short-term matings probably do not experience difficulty obtaining willing sexual partners. Women in a long-term relationship may present the highest risk of sperm competition. The primary partner's frequent inseminations might therefore make women in a long-term relationship least attractive as short-term sexual partners.

As predicted, Shackelford *et al.* (2004) found that men's reported likelihood of pursuing a short-term sexual relationship was lowest when imagining that the potential short-term partner is married, next lowest when imagining that she is not married but involved in casual sexual relationships, and highest when imagining that she is not married and not involved in any casual sexual relationships. These results suggest that, when selecting short-term sexual partners, men do so in part to avoid sperm competition.

An alternative explanation for the pattern of results is that by preferring unmated women, men can avoiding the costs associated with contracting a sexually transmitted disease (STD). The data, however, refute this alternative explanation. The potential short-term partner most likely to be infected with an STD would be the one having casual sex and, therefore, would be least preferred according to this alternative hypothesis. The married potential sexual partner, however, was the least preferred. Men's preferences, therefore, suggest that avoiding STDs may be less important than avoiding sperm competition when selecting short-term partners.

SPERM COMPETITION AND MEN'S SEXUAL AROUSAL AND SEXUAL FANTASIES

It is well documented that men's sexual fantasies often involve multiple, anonymous partners (Ellis & Symons, 1990), but men's sexual fantasies include more than sexual variety. Because sperm competition seems to have been a recurrent feature of human evolutionary history, it may be useful to interpret some facets of men's sexual fantasies in the light of sperm competition.

Although never investigated empirically, one may assert with confidence that many men are sexually aroused by the exclusive sexual interaction between two women. Hollywood seems to be aware of this preference as well. A common scenario in many mainstream movies and television shows, for example, involves two women (often implied or explicit heterosexuals) kissing or performing other sexual acts with one another while an audience of one or more men observes the acts and becomes sexually aroused. Similarly, two women dancing seductively with one another tends to stimulate interest among observing men. It could be argued that the sight of two heterosexual women engaging in sexual behaviors is sexually arousing because it suggests both women are sexually available and copulation with both is imminent. An interpretation informed by sperm-competition theory, however, might argue that the sight of two heterosexual women engaging in sexual behaviors is sexually arousing because it is a cue to an absence of sperm competition. If given a choice, men might prefer to avoid sperm competition and thus be the sole fertilizer of a woman's eggs. Thus two women engaging in sexual behaviors may signal to men that the women are without male partners and, therefore, pose no risk of sperm competition. Although highly speculative and difficult to test, this hypothesis serves to illustrate how the application of sperm competition to human mating psychology and behavior generates interesting and novel hypotheses.

Although the absence of sperm competition in a potential sexual partner is expected to be sexually arousing, it also has been argued that the *presence* of sperm competition may result in sexual arousal. Pound (2002) argued that men should find cues of increased sperm-competition risk to be sexually arousing because frequent copulation can be an effective method of paternity assurance. Pound (2002) hypothesized that men, therefore, should be more aroused by pornography that incorporates cues of sperm competition than by comparable material in which such cues are absent. Content analyses of pornographic images on world wide web sites and of commercial "adult" video releases revealed that depictions of sexual activity involving a female and multiple males are more prevalent than those involving a male and multiple females. An online survey of self-reported preferences and an online preference study that unobtrusively examined image-selection behavior yielded corroborative results. Pound (2002) argued that the most parsimonious explanation for such results is that male arousal in response to visual cues of sperm-competition risk reflects the functioning of psychological mechanisms that would have motivated adaptive patterns of copulatory behavior in ancestral males exposed to evidence of female promiscuity.

The idea that men might experience increased sexual motivation in response to cues of sperm-competition risk is also supported by anecdotal accounts of men who engage in "swinging" or "partner-swapping." Encouraging one's partner to copulate with other men is obviously a maladaptive strategy in that it clearly increases the risk of cuckoldry. However, it seems that in some contemporary societies some men do just this – perhaps because such men often report that they find the sight of their partner interacting sexually with other men to be sexually arousing (Talese, 1981). Moreover, they report that they experience increased sexual desire for their partner following her sexual encounters with other men, and some indicate that this increase in desire is particularly acute when they have witnessed their partner having sexual intercourse with another man (Gould, 1999).

Men may also voluntarily expose themselves to cues of sperm-competition risk through their participation in sexual "role-playing" with their partner. Pretending to be someone other than himself may activate mechanisms in men associated with an increased risk of sperm competition, resulting in increased sexual arousal. For example, by "role-playing" a man might get to hear his partner talk as if she were copulating with another man. Alternatively, role-playing may be sexually arousing to men and women because it is exploiting mechanisms associated with sexual variety. Teasing the two hypotheses apart would require, among other tests, documenting how willing or excited men and women are to adopt a different role during role-playing. If the data revealed that when role-playing with their partners men are willing and excited to adopt a different role themselves, while simultaneously unconcerned with whether or not their female partners adopts a different role, this may constitute preliminary support for the sperm-competition risk hypothesis. Again, applying sperm-competition theory to aspects of human sexual psychology and behavior may generate unique perspectives and hypotheses.

IS THERE EVIDENCE OF CONTEST COMPETITION BETWEEN MEN'S EJACULATES?

Apart from the remarkable feat of traversing a hostile reproductive tract to fertilize an ovum or ova, sperm do some astonishing things. Sperm of the common wood mouse (*Apodemus sylvaticus*) have a hook that allows the sperm to adhere to one another to form a motile "train" of several thousand sperm (Moore *et al.*, 2002). These trains display greater motility and velocity than single sperm, facilitating fertilization. This cooperative behavior between sperm of a single male reveals that sperm are capable of complex behavior. Might mammalian sperm display equally complex behavior in the presence of rival sperm?

Baker and Bellis (1988) proposed that, in mammals, postcopulatory competition between rival male ejaculates might involve more that just scramble competition and that rival sperm may interfere actively with each other's ability to fertilize ova. Mammalian ejaculates contain sperm that are polymorphic (i.e. existing in different morphologies or shapes and sizes). Previously interpreted as the result of developmental error (Cohen, 1973), Baker and Bellis (1988) proposed that sperm polymorphism was not due to meiotic errors, but instead reflected a functionally adaptive "division of labor" between sperm. Baker and Bellis (1988) proposed two categories of sperm: "egg-getters" and "kamikaze" sperm. Egg-getters comprise the small proportion of sperm programmed to fertilize ova. Baker and Bellis (1988) argued that most of the ejaculate is composed of kamikaze sperm that function to prevent other males' sperm from fertilizing the ova by forming a barrier at strategic positions within the reproductive tract. Preliminary evidence for Baker and Bellis' (1988) Kamikaze Sperm Hypothesis came from the observation that the copulatory plugs of bats are composed of so-called "malformed" sperm (Fenton, 1984), and from documentation that, in laboratory mice, different proportions of sperm morphs are found reliably at particular positions within the female reproductive tract (Cohen, 1977).

Harcourt (1989) challenged the Kamikaze Sperm Hypothesis. Harcourt argued that "malformed" sperm were unlikely to have adaptive functions, citing evidence from Wildt *et al.* (1987) that, in lions, inbreeding results in an increase in the proportion of deformed sperm. Harcourt (1989) argued that, if deformed sperm were produced by an adaptation, inbreeding would not increase the expression of the trait, but instead would decrease it. Harcourt (1989) also argued that the presence of malformed sperm in the copulatory plugs of bats is a consequence of the malformed sperm's poor mobility and, therefore, that plug formation was not a designed function of deformed sperm. Following Cohen (1973), Harcourt (1989, p. 864) concluded that "abnormal sperm are still best explained by errors in production."

Baker and Bellis (1989b) responded to Harcourt's (1989) objections and elaborated on the Kamikaze Sperm Hypothesis. In their elaboration, Baker and Bellis (1989b) proposed a more active role for kamikaze sperm, speculating that evolutionary arms races between ejaculates could result in kamikaze sperm that incapacitate rival sperm with acrosomal enzymes or by inducing attack by female leukocytes. Baker and Bellis (1995) proposed specialized roles for kamikaze sperm and identified two categories of kamikaze sperm: "blockers" and "seek-and-destroyers." Baker and Bellis (1995) documented that, when mixing ejaculates from two different men *in vitro*, agglutination and mortality of sperm increased. Baker and Bellis (1995) interpreted these findings as an indication that, when encountering sperm from another male, some sperm impede the progress of rival sperm (blockers) and some sperm attack and incapacitate rival sperm (seek-and-destroyers). The Kamikaze Sperm Hypothesis and the reported interaction of rival sperm have generated substantial criticism, however (see, e.g. Birkhead, Moore, & Bedford, 1997; Short, 1998).

Moore, Martin, and Birkhead (1999) performed the first and, thus far, only attempt to replicate some of Baker and Bellis' (1995) work, but failed to replicate the findings of Baker and Bellis (1995). It should be noted, however, that only a few of the predictions derived from the Kamikaze Sperm Hypothesis were tested by Baker and Bellis (1995) and even fewer were tested by Moore *et al.* (1999). After mixing sperm from different men and comparing these heterospermic samples to self-sperm (i.e. homospermic) samples, Moore *et al.* (1999) observed no increase in aggregation and no greater incidence of incapacitated sperm in the heterospermic samples. Moore *et al.* (1999) did not replicate exactly the methodological procedures used by Baker and Bellis (1995), however. Heterospermic and homospermic samples, for example, were allowed to interact for just 1–3 h, whereas Baker and Bellis (1995) allowed them to interact for fully 3–6 h. Moore *et al.* (1999) offered theoretical reasons for this shorter interactive window (i.e. because 1–3 h is the time that sperm normally remain in the human vagina), but perhaps this interval was too restrictive. Upon insemination, sperm have one of two initial fates: some are ejected or secreted from the vagina and some travel quickly from the vagina to the cervix and uterus. Perhaps the majority of sperm warfare takes place in the cervix and uterus, locations in the reproductive tract where sperm are able to interact for a prolonged period. If this is the case, Baker and Bellis' (1995) longer, 3–6 h interactive window would be more valid ecologically. In addition, both Baker and Bellis (1995) and Moore *et al.* (1999) investigated sperm interactions *in vitro*, and one cannot be sure that sperm in a petri dish behave precisely as they do in the human vagina.

Aside from Moore *et al.*'s (1999) failure to replicate Baker and Bellis' (1995) findings, additional skepticism is generated by Baker and Bellis' (1995) failure to clearly specify how sperm can differentiate self-sperm from non-self-sperm. Given that sperm consist of a diminutive single-cell structure, a self-recognition system that must differentiate between not just different genes (because even sperm from a single male contain different combinations of genes), but different *sets* of competing genes (i.e. genes from another male) may be unlikely to have evolved. Moore *et al.*'s (1999) failure to replicate Baker and Bellis' (1995) findings and the absence of a clear self-recognition system is not fatal to the Kamikaze Sperm Hypothesis, but such concerns are cause for skepticism about its plausibility. Clearly, more work remains before we can draw a clear conclusion about the status of the hypothesis. Recent work by Kura and Nakashima (2000) might be viewed as encouraging for supporters of the hypothesis, however. Kura and Nakashima (2000) used theoretical and mathematical models to describe the conditions necessary for soldier sperm classes to evolve. Kura and Nakashima (2000) concluded that such conditions are not stringent and far from unlikely.

Concluding remarks

This chapter reviews the mechanism of postcopulatory sexual selection first identified by Geoff Parker (1970): sperm competition. Sperm competition and its effects have been documented or inferred to exist in dozens of non-human species, but researchers are beginning to uncover adaptations in humans that are most parsimoniously explained by sperm-competition theory. In humans, sperm-competition may have influenced reproductive anatomy and physiology, men's attraction to and sexual interest in their partners, men's copulatory behaviors, men's short-term mate selection, and men's sexual arousal and sexual fantasies.

Although this chapter focuses on men's adaptations to sperm competition, women are not simply passive sperm receptacles. If sperm competition was a recurrent feature of human evolutionary history, we would expect to identify adaptations not only in men but also in women. Indeed, intersexual conflict between ancestral males and females produces a coevolutionary arms race between the sexes, in which an advantage gained by one sex selects for counter-adaptations in the other sex (see, e.g. Rice, 1996). Thus, men's numerous adaptations to sperm competition are likely to be met by numerous adaptations in women. Women's adaptations to sperm competition are considered in Shackelford *et al.* (2005).

References

Arndt, W. B., Jr., Foehl, J. C., and Good, F. E. (1985). Specific sexual fantasy themes: a multidimensional study. *Journal of Personality and Social Psychology*, **48**, 472–80.

Baker, R. R. and Bellis, M. A. (1988). "Kamikaze" sperm in mammals? *Animal Behaviour*, **36**, 936–9.

Baker, R. R. and Bellis, M. A. (1989a). Number of sperm in human ejaculates varies in accordance with sperm competition theory. *Animal Behaviour*, **37**, 867–9.

Baker, R. R. and Bellis, M. A. (1989b). Elaboration of the kamikaze sperm hypothesis: a reply to Harcourt. *Animal Behaviour*, **37**, 865–7.

Baker, R. R. and Bellis, M. A. (1993). Human sperm competition: ejaculate adjustment by males and the function of masturbation. *Animal Behaviour*, **46**, 861–85.

Baker, R. R. and Bellis, M. A. (1995). *Human Sperm Competition*. London: Chapman & Hall.

Bellis, M. A. and Baker, R. R. (1990). Do females promote sperm competition: data for humans. *Animal Behavior*, **40**, 197–9.

Bellis, M. A., Baker, R. R., and Gage, M. J. G. (1990). Variation in rat ejaculates consistent with the Kamikaze Sperm Hypothesis. *Journal of Mammalogy*, **71**, 479–80.

Birkhead, T. R. (2000). *Promiscuity*. London: Faber and Faber.

Birkhead, T. R. and Møller, A. P. (1992). *Sperm Competition in Birds*. London: Academic Press.

Birkhead, T. R. and Møller, A. P. (1998). *Sperm Competition and Sexual Selection*. New York: Academic Press.

Birkhead, T. R., Moore, H. D. M., and Bedford, J. M. (1997). Sex, science, and sensationalism. *Trends in Ecology and Evolution*, **12**, 121–2.

Buss, D. M. (2004). *The Evolution of Desire* (2nd edn.). New York: Basic Books.

Buss, D. M. and Schmitt, D. P. (1993). Sexual strategies theory: an evolutionary perspective on human mating. *Psychological Review*, **100**, 204–32.

Buss, D. M., Larsen, R. J., Westen, D., and Semmelroth, J. (1992). Sex differences in jealousy: evolution, physiology and psychology. *Psychological Science*, **3**, 251–5.

Chien, L. (2003). Does quality of marital sex decline with duration? *Archives of Sexual Behavior*, **32**, 55–60.

Cohen, J. (1973). Cross-overs, sperm redundancy and their close association. *Heredity*, **31**, 408–13.

Cohen, J. (1977). *Reproduction*. London: Butterworth.

Cook, P. A. and Wedell, N. (1996). Ejaculate dynamics in butterflies: a strategy for maximizing fertilization success? *Proceedings of the Royal Society of London B*, **263**, 1047–51.

Daly, M., Wilson, M., and Weghorst, J. (1982). Male sexual jealousy. *Ethology and Sociobiology*, **3**, 11–27.

Darwin, C. (1871). *The Descent of Man and Selection in Relation to Sex*. London: Murray.

Dixson, A. F. (1998). *Primate Sexuality*. Oxford: Oxford University Press.

Eberhard, W. G. (1996). *Female Control*. Princeton, NJ: Princeton University Press.

Ellis, B. J. and Symons, D. (1990). Sex differences in sexual fantasy: an evolutionary psychological approach. *Journal of Sex Research*, **27**, 527–55.

Fenton, M. B. (1984). The case of vepertilionid and rhinolophid bats. In R. L. Smith, ed., *Sperm Competition and the Evolution of Animal Mating Systems*. London: Academic Press, pp. 573–87.

Gage, A. R. and Barnard, C. J. (1996). Male crickets increase sperm number in relation to competition and female size. *Behavioral Ecology and Sociobiology*, **38**, 349–53.

Gage, M. J. G. (1994). Associations between body-size, mating pattern, testis size and sperm lengths across butterflies. *Proceedings of the Royal Society of London B*, **258**, 247–54.

Gallup, G. G. and Burch, R. L. (2004). Semen displacement as a sperm competition strategy in humans. *Evolutionary Psychology*, **2**, 12–23.

Gallup, G. G., Burch, R. L., Zappieri, M. L., Parvez, R. A., Stockwell, M. L., and Davis, J. A. (2003). The human penis as a semen displacement device. *Evolution and Human Behavior*, **24**, 277–89.

Gangestad, S. W. and Simpson, J. A. (2000). The evolution of human mating: trade-offs and strategic pluralism. *Behavior and Brain Sciences*, **23**, 573–87.

Ginsberg, J. R. and Rubenstein, D. I. (1990). Sperm competition and variation in zebra mating behavior. *Behavioral Ecology and Sociobiology*, **26**, 427–34.

Goetz, A. T., Shackelford, T. K., Weekes-Shackelford, V. A., *et al.* (2005). Mate retention, semen displacement, and human sperm competition: a preliminary investigation of tactics to prevent and correct female infidelity. *Personality and Individual Differences*, **38**, 749–63.

Gomendio, M. and Roldán, E. R. S. (1993). Mechanisms of sperm competition: linking physiology and behavioral ecology. *Trends in Ecology and Evolution*, **8**, 95–100.

Gomendio, M., Harcourt, A. H., and Roldán, E. R. S. (1998). Sperm competition in mammals. In T. R. Birkhead and A. P. Møller, eds., *Sperm Competition and Sexual Selection*. New York: Academic Press, pp. 667–756.

Gould, T. (1999). *The Lifestyle*. New York: Firefly.

Greiling, H. and Buss, D. M. (2000). Women's sexual strategies: the hidden dimension of extra-pair mating. *Personality and Individual Differences*, **28**, 929–63.

Harcourt, A. H. (1989). Deformed sperm are probably not adaptive. *Animal Behaviour*, **37**, 863–5.

Harcourt, A. H., Harvey, P. H., Larson, S. G., and Short, R. V. (1981). Testis weight, body weight, and breeding system in primates. *Nature*, **293**, 55–7.

Harvey, P. H. and Harcourt, A. H. (1984). Sperm competition, testis size, and breeding systems in primates. In R. L. Smith, ed., *Sperm Competition and the Evolution of Animal Mating Systems*. San Diego: Academic Press, pp. 589–600.

Hosken, D. J. and Ward, P. I. (2001). Experimental evidence for testis size evolution via sperm competition. *Ecology Letters*, **4**, 10–13.

Hrdy, S. B. (1981). *The Woman that Never Evolved*. Cambridge: Harvard University Press.

Hunt, M. (1974). *Sexual Behavior in the 70s*. Chicago: Playboy Press.

Jennions, M. D. and Passmore, N. I. (1993). Sperm competition in frogs: testis size and a sterile male experiment on *Chiromantis-xerampelina (Rhacophoridae)*. *Biological Journal of the Linnean Society*, **50**, 211–20.

Jennions, M. D. and Petrie, M. (2000). Why do females mate multiply? A review of the genetic benefits. *Biological Reviews*, **75**, 21–64.

Johnson, A. M., Mercer, C. H., Erens, B., *et al.* (2001). Sexual behaviour in Britain: partnerships, practices, and HIV risk behaviours. *Lancet*, **358**, 1835–42.

Klusmann, D. (2002). Sexual motivation and the duration of partnership. *Archives of Sexual Behavior*, **31**, 275–87.

Kura, T. and Nakashima, Y. (2000). Conditions for the evolution of soldier sperm classes. *Evolution*, **54**, 72–80.

Laumann, E. O., Gagnon, J. H., Michael, R. T., and Michaels, S. (1994). *The Social Organization of Sexuality*. Chicago: University of Chicago Press.

Leitenberg, H. and Henning, K. (1995). Sexual fantasy. *Psychological Bulletin*, **117**, 469–96.

Møller, A. P. (1985). Mixed reproductive strategy and mate guarding in a semi-colonial passerine, the swallow *Hirundo rustica*. *Behavioral Ecology and Sociobiology*, **17**, 401–8.

Møller, A. P. (1987). Mate guarding in the swallow *Hirundo rustica*. *Behavioral Ecology and Sociobiology*, **21**, 119–23.

Møller, A. P. (1988a). Testes size, ejaculate quality and sperm competition in birds. *Biological Journal of the Linnean Society*, **33**, 273–83.

Møller, A. P. (1988b). Paternity and paternal care in the swallow, *Hirundo rustica*. *Animal Behaviour*, **36**, 996–1005.

Moore, H. D., Martin, M., and Birkhead, T. R. (1999). No evidence for killer sperm or other selective interactions between human spermatozoa in ejaculates of different males in vitro. *Proceedings of the Royal Society of London B*, **266**, 2343–50.

Moore, H. D., Dvorakova, K., Jenkins, N., and Breed, W. (2002). Exceptional sperm cooperation in the wood mouse. *Nature*, **418**, 174–7.

Parker, G. A. (1970). Sperm competition and its evolutionary consequences in the insects. *Biological Reviews*, **45**, 525–67.

Parker, G. A. (1982). Why are there so many tiny sperm? Sperm competition and the maintenance of two sexes. *Journal of Theoretical Biology*, **96**, 281–94.

Parker, G. A. (1990a). Sperm competition games: raffles and roles. *Proceedings of the Royal Society of London B*, **242**, 120–6.

Parker, G. A. (1990b). Sperm competition games: sneaks and extra-pair copulations. *Proceedings of the Royal Society of London Series B*, **242**, 127–33.

Parker, G. A., Ball, M. A., Stockley, P., and Gage, M. J. G. (1997). Sperm competition games: a prospective analysis of risk assessment. *Proceedings of the Royal Society of London B*, **264**, 1793–802.

Pelletier, L. A. and Herold, E. S. (1988). The relationship of age, sex guilt, and sexual experience with female sexual fantasies. *Journal of Sex Research*, **24**, 250–6.

Person, E. S., Terestman, N., Myers, W. A., Goldberg, E. L., and Salvadori, C. (1989). Gender differences in sexual behaviors and fantasies in a college population. *Journal of Sex and Marital Therapy*, **15**, 187–98.

Platek, S. M. (2003). Effects of paternal resemblance on paternal investment: an evolutionary model. *Evolution and Cognition*, **9**, 189–97.

Pound, N. (2002). Male interest in visual cues of sperm competition risk. *Evolution and Human Behavior*, **23**, 443–66.

Pound, N. and Gage. M. J. G. (2004). Prudent sperm allocation in *Rattus norvegicus*: a mammalian model of adaptive ejaculate adjustment. *Animal Behaviour*, **68**, 819–23.

Pound, N., Javed, M. H., Ruberto, C., Shaikh, M. A., and Del Valle, A. P. (2002). Duration of sexual arousal predicts semen parameters for masturbatory ejaculates. *Physiology and Behavior*, **76**, 685–9.

Price, J. H. and Miller, P. A. (1984). Sexual fantasies of Black and of White college students. *Psychological Reports*, **54**, 1007–14.

Rice, W. R. (1996). Sexually antagonistic male adaptation triggered by experimental arrest of female evolution. *Nature*, **381**, 232–4.

Rokach, A. (1990). Content analysis of sexual fantasies of males and females. *Journal of Psychology*, **124**, 427–36.

Schmitt, D. P., Shackelford, T. K., and Buss, D. M. (2001). Are men really more oriented toward short-term mating than women? A critical review of theory and research. *Psychology, Evolution, and Gender*, **3**, 211–39.

Schmitt, D. P., Shackelford, T. K., Duntley, J., Tooke, W., and Buss, D. M. (2001). The desire for sexual variety as a tool for understanding basic human mating strategies. *Personal Relationships*, **8**, 425–55.

Schmitt, D. P., Alcalay, L., Allik, J., et al. (2003). Universal sex differences in the desire for sexual variety: tests from 52 nations, 6 continents, and 13 islands. *Journal of Personality and Social Psychology*, **85**, 85–104.

Shackelford, T. K. (2003). Preventing, correcting, and anticipating female infidelity: three adaptive problems of sperm competition. *Evolution and Cognition*, **9**, 90–6.

Shackelford, T. K. and LeBlanc, G. J. (2001). Sperm competition in insects, birds, and humans: insights from a comparative evolutionary perspective. *Evolution and Cognition*, **7**, 194–202.

Shackelford, T. K., Weekes-Shackelford, V. A., LeBlanc, G. J., et al. (2000). Female coital orgasm and male attractiveness. *Human Nature*, **11**, 299–306.

Shackelford, T. K., LeBlanc, G. J., Weekes-Shackelford, V. A., et al. (2002). Psychological adaptation to human sperm competition. *Evolution and Human Behavior*, **23**, 123–38.

Shackelford, T. K., Goetz, A. T., LaMunyon, C. W., Quintus, B. J., and Weekes-Shackelford, V. A. (2004). Sex differences in sexual psychology produce sex similar preferences for a short-term mate. *Archives of Sexual Behavior*, **33**, 405–12.

Shackelford, T. K., Pound, N., and Goetz, A. T. (2005). Psychological and physiological adaptation to human sperm competition. *Review of General Psychology*, **9**, 228–48.

Short, R. V. (1979). Sexual selection and its component parts, somatic and genital selection as illustrated by man and the great apes. *Advances in the Study of Behavior*, **9**, 131–58.

Short, R. V. (1981). Sexual selection in man and the great apes. In C. E. Graham, ed., *Reproductive Biology of the Great Apes*. New York: Academic Press, pp. 319–41.

Short, R. V. (1998). Review of Human Sperm Competition: Copulation, Masturbation and Infidelity, by R. R. Baker and M. A. Bellis. *European Sociobiology Society Newsletter*, **47**, 20–3.

Simmons, L. W. (2001). *Sperm Competition and its Evolutionary Consequences in the Insects*. Princeton, NJ: Princeton University Press.

Smith, R. L. (1984). Human sperm competition. In R. L. Smith, ed., *Sperm Competition and the Evolution of Animal Mating Systems*. New York: Academic Press, pp. 601–60.

Symons, D. (1979). *The Evolution of Human Sexuality*. New York: Oxford University Press.

Talese, G. (1981). *Thy Neighbor's Wife*. New York: Ballantine.

Thornhill, R. (1983). Cryptic female choice and its implications in the scorpionfly *Harpobittacus nigriceps*. *American Naturalist*, **122**, 765–88.

Trivers, R. L. (1972). Parental investment and sexual selection. In B. Campbell, ed., *Sexual Selection and the Descent of Man*. London: Aldine, pp. 139–79.

Wedell, N., Gage, M. J. G., and Parker, G. A. (2002). Sperm competition, male prudence and sperm-limited females. *Trends in Ecology and Evolution*, **17**, 313–20.

Wildt, D. E., Bush, M., Goodrowe, K. L., et al. (1987). Reproductive and genetic consequences of founding isolated lion populations. *Nature*, **329**, 328–31.

Wilson, G. D. (1987). Male-female differences in sexual activity, enjoyment and fantasies. *Personality and Individual Differences*, **8**, 125–7.

Wilson, G. D. (1997). Gender differences in sexual fantasy: an evolutionary analysis. *Personality and Individual Differences*, **22**, 27–31.

Wilson, G. D. and Lang, R. J. (1981). Sex differences in sexual fantasy patterns. *Personality and Individual Differences*, **2**, 343–6.

Wilson, M. and Daly, M. (1992). The man who mistook his wife for a chattel. In J. H. Barkow, L. Cosmides, and J. Tooby, eds., *The Adapted Mind*. New York: Oxford University Press, pp. 19–136.

Wyckoff, G. J., Wang, W., and Wu, C. (2000). Rapid evolution of male reproductive genes in the descent of man. *Nature*, **403**, 304–8.

Zavos, P. M. (1985). Seminal parameters of ejaculates collected from oligospermic and normospermic patients via masturbation and at intercourse with the use of a Silastic seminal fluid collection device. *Fertility and Sterility*, **44**, 517–20.

Zavos, P. M. (1988). Seminal parameters of ejaculates collected at intercourse with the use of a seminal collection device with different levels of precoital stimulation. *Journal of Andrology*, **9**, P-36.

Zavos, P. M. and Goodpasture, J. C. (1989). Clinical improvements of specific seminal deficiencies via intercourse with a seminal collection device versus masturbation. *Fertility and Sterility*, **51**, 190–3.

Zavos, P. M. Kofinas, G. D., Sofikitis, N. V., Zarmakoupis, P. N., and Miyagawa, I. (1994). Differences in seminal parameters in specimens collected via intercourse and incomplete intercourse (coitus interruptus). *Fertility and Sterility*, **61**, 1174–6.

7

The semen-displacement hypothesis: semen hydraulics and the intra-pair copulation proclivity model of female infidelity

GORDON G. GALLUP, JR.
SUNY at Albany
AND
REBECCA L. BURCH
SUNY at Oswego

Introduction

Among sexually reproducing species, the penis evolved as an internal fertilization device. But across different species, penises exist in a bewildering array of shapes and sizes (see Eberhard, 1985). Among primates, the human penis is distinctive by virtue of both its size and its enlarged glans and protruding coronal ridge (see Gallup & Burch, 2004). There has been some speculation that the human penis evolved not only as an internal fertilization device, but also as a mechanism for displacing semen left by rival males in the female reproductive tract (e.g. Baker & Bellis, 1995).

In a series of studies designed to simulate sexual intercourse under laboratory conditions using artificial genitals, we found that when latex vaginas contained simulated semen, phalluses that approximated the configuration of the human penis displaced 80% or more of the semen by drawing it away from the cervical end of the vagina (Gallup *et al.*, 2003). Through a series of experimental manipulations, we determined that the coronal ridge may be an important feature of the penis in mediating semen displacement. Thus, as a mechanical means of affecting sperm competition, the human penis may enable successive males to displace foreign semen from the female reproductive tract and substitute their semen for those of their rivals.

In a survey of over 600 college students, reported in the same paper (Gallup *et al.*, 2003), we found that, following periods of separation or allegations of female infidelity, both males and females report that penile thrusting is noticeably deeper and more vigorous. Therefore, how men unwittingly use their penis also may be related to its semen-displacement properties, and the parameters of penile thrusting appear to vary as a function of the likelihood that there may be semen from other males in the female reproductive tract.

In a more recent paper, we derived and expanded on a number of other theoretical implications that follow from the proposition that the unique configuration of the human penis evolved to compete with semen left in the vagina by other males (Gallup & Burch, 2004). Included among the predictions we made are the following. If several males copulate with the same female in a short period of time (e.g. 24 h), the last male to inseminate the female has the best chance of paternity (referred to as ordinal ejaculation effects or last-male precedence; Gallup & Burch, 2004). Because of the possibility that males may unwittingly piggyback foreign semen from one vagina to the next, females that copulate with uncircumcised partners are at risk of fertilization by proxy and resulting impregnation by males they have never had sex with. Likewise, we argued that, in one sense, premature ejaculation can be thought of as a failure to achieve semen displacement, and we speculated that jealousy-induction procedures might work to antagonize this problem. Finally, because of the prospect of displacing their own semen, we predicted that males who show vigorous post-ejaculatory thrusting are at risk of infertility as a consequence of displacing their own semen.

In this chapter, we expand on some of these and other implications of the semen-displacement hypothesis. We also present preliminary data from several surveys we have conducted that bear on some of the predictions that follow from this hypothesis.

Semen hydraulics

One way to think about semen displacement is in terms of semen hydraulics, or the flow and resistance of seminal fluid as it passes through confined spaces. For our purposes, there are three applications of semen hydraulics that relate to human reproductive competition: semen viscosity, semen displacement, and semen retention.

One of the principal parameters of semen hydraulics is *semen viscosity*. There is considerable individual variation among men in semen viscosity (Gonzales, Kortebani, & Mazzolli, 1993), and the viscosity of semen has been implicated as a factor in male fertility. Semen hyperviscosity has been linked to infertility and is often one of the initial assessments performed by infertility clinics (Gonzales, Kortebani, & Mazzolli, 1993; Gopalkrishnan, Padwal, & Balaiah, 2000). Thick or viscous semen appears to entrap sperm and impair their ability to swim up through the female reproductive tract.

There also is a relationship between semen viscosity and *semen displacement*. In our attempt to model semen displacement under laboratory conditions, we examined the effect of variation in simulated semen viscosity (Gallup *et al.*, 2003). Using different levels of semen viscosity and several different phalluses, we determined that the magnitude of semen displacement was inversely proportional to semen viscosity; that is, viscous semen was less subject to displacement. Thus, there would appear to be a trade-off between semen viscosity and semen displacement. Viscous semen deposited deep inside the vaginal tract would be less accessible and more difficult for other males to displace. However, because semen hyperviscosity compromises sperm motility, there is probably a trade-off between semen displacement and fertility with respect to semen viscosity; that is, the benefits of semen viscosity as a means of minimizing displacement are offset by the fertility costs that are incurred as a consequence.

The same constraints hold true for *semen retention*. With the assumption of an upright posture during human evolution and the emergence of patterns of bipedal locomotion, the female reproductive tract has been brought into a perpendicular orientation with gravity and as such is poorly suited to the retention of semen (Gallup & Suarez, 1983). In most quadruped species (including most primates), copulation occurs in the dorsal–ventral mode with the male mounting the female from behind, and under these conditions the female reproductive tract is in a parallel orientation to gravity. The problems posed by semen retention in bipedal humans appear to have impacted human sexual behavior. One of the most common, cross-cultural means of affecting copulation in humans is in face-to-face or frontal encounters that occur in the ventral–ventral mode, with the female in a supine position and the male on top (Ford & Beach, 1951). In this position the woman's reproductive tract is brought back into a more primitive, parallel orientation with gravity and as such may be better suited to semen retention. Because the orientation of the female reproductive tract may be important to the retention of semen, we predict that couples that frequently use the female superior position will be at greater risk of experiencing infertility problems.

It follows from this analysis that we expect to find a variety of corollary postcopulatory adaptations in humans that function to postpone the assumption of an upright posture following insemination. With regard to the timing of copulation, humans tend to show a nocturnal copulation bias (Ford & Beach,

1951). By copulating preferentially at night, the likelihood of remaining in a prone position for an extended period of time afterward is enhanced. The sedative-like effects of orgasm would further promote remaining in a prone position following a sexual encounter (Gallup & Suarez, 1983), particularly under conditions in which sexual partners sleep together (Hughes, Harrison, & Gallup, 2004). The fact that some human females are capable of experiencing multiple orgasms may cause these sedative-like effects to summate and further promote remaining in a supine position for an extended period of time after sex.

Postcopulatory behavior

In addition to the premium that would accrue to depositing semen in the most inaccessible parts of the vagina and thereby minimizing the likelihood of displacement by rival males, the same logic holds for semen retention. Semen deposited deep within the vaginal tract would have a retention advantage, and this may have produced additional selective pressure for the evolution of a longer penis (see Gallup & Burch, 2004; Gallup & Suarez, 1983). As implied by our discussion of semen hydraulics, another important parameter of semen retention involves postcopulatory behavior. What females do after they copulate can have important implications for semen retention.

As one test of this hypothesis, we examined differences in postcopulatory behavior among people from different races (Stockwell et al., 2001). There are substantial differences in penis size across different racial groups, with African-Americans having, on average, longer penises (e.g. Gebhard & Johnson, 1979). To see if these differences evolved to compensate for differences in what females do after they copulate, we surveyed over 500 undergraduates regarding their ethnic background, sexual practices, and the behavior of their partners. Consistent with the hypothesis that semen deposited deep within the vaginal tract would be afforded a retention advantage, African-American women were more likely than Caucasian and Asian women to get up and do other things right after having sexual intercourse. After sex, African-American women were more likely than women of other ethnic backgrounds to report taking a shower, getting dressed, going home, or going to work. They were also less likely to remain in a prone position and to cuddle with their partners after sex.

In light of the dual function of the penis as both a means of maximizing semen retention and displacing semen from rival males, it is notable that there are independent data implicating higher levels of female infidelity among African-Americans based on the existence of ethnic differences in jealousy and partner abuse. In a recent study by McFarlane and Wilson (2000), the majority of victims of domestic violence were African-American. Likewise, Straus and Gelles

(1986) have shown that rates of overall violence against females are higher among African-American men, and family violence that ends in homicide is higher as well (Wilson & Daly, 1992). In a sample of 780 college students, we also found significant race differences in domestic violence (Stockwell *et al.*, 2000), with a higher incidence of abusive acts directed toward female partners among African-Americans than women of other ethnic groups.

Double-mating

The semen-displacement hypothesis assumes that there were recurrent situations during human evolutionary history in which females had sex with two or more males in fairly close temporal proximity to one another. We listed the following situations as ones that satisfy this criterion: (1) consensual sex with multiple concurrent partners, (2) nonconsensual sex with multiple concurrent partners, and (3) successive consensual or nonconsensual sexual encounters with multiple partners that occur within a relatively brief period of time (Gallup & Burch, 2004). Contemporary examples of female double-mating include threesomes, group sex, gang rape, extra-pair copulations, promiscuity, and prostitution.

Using the definition of double-mating as a woman having sexual intercourse with a man while another man's ejaculate is still present in her reproductive tract, Baker and Bellis (1995) surveyed over 3500 women through a British magazine. They found that the lifetime number of sexual partners varied considerably; after 3000 copulations, the percentage of women who had over 100 partners was the same as those who only had one (5.6%). They also found that as women age, and as they have more children, the incidence of extra-pair copulations increases.

The incidence of double-mating in Baker and Bellis' British sample also depended upon sexual experience; roughly one out of six women double-mated within their first 50 copulations, one-half double-mated within their first 500 copulations, and over 80% had double-mated by the time they had experienced 3000 copulations. After 500 copulations, one in 200 women reported being inseminated on at least one occasion by two different men within 30 min of one another. Approximately 30% of women in the Baker and Bellis sample reported having been inseminated by two males within 24 h.

In a recent attempt to examine the incidence of double-mating in a contemporary sample of college students, we found that 18%, or over one out of six of the female respondents ($n = 136$), and 10% of the males ($n = 44$) reported having had at least one extra-pair copulation while being in a committed relationship (Gallup, Burch, and Beren Mitchell, 2005). In terms of sex with successive partners, one out of 10 (10.4%) females acknowledged having had intercourse with more than one male in a 24-h period, 3.5% reported having sex with more than one male in 12 h, and 2.6% admitted to having sex with two or more men within 8 h or less. In response to questions about concurrent (rather than successive) encounters with multiple sex partners, 8.1%, or one out of 12 females, reported having engaged in a threesome, and 5.9% of the females acknowledged having participated in group sex with three or more partners. These data suggest that the incidence of double-mating among female college students is substantial. Moreover, these data should be treated as conservative estimates of double-mating. First, even under conditions of anonymity, women have little to gain (and far more to lose) by being completely candid and forthright about their sexual indiscretions. Second, college students in their late teens and early twenties are relatively inexperienced in the sexual domain as compared to older, more mature women.

Adaptations to self-semen displacement

If the human penis evolved to displace semen left by other males, what is to prevent this adaptation from displacing the male's own semen? The data derived from artificial genitals (Gallup *et al.*, 2003) suggest that continued thrusting beyond the point of ejaculation would lead to displacement of the male's own semen. Therefore, the tenability of the displacement hypothesis depends on identifying evolved collateral mechanisms that serve to minimize post-ejaculatory thrusting and thereby reduce the likelihood of self-semen displacement.

Candidate mechanisms identified by Gallup and Burch (2004) that appear to preclude or at least diminish self-semen displacement include the following post-ejaculatory changes: (1) penile hypersensitivity, (2) loss of an erection, and (3) the refractory period. Due to enhanced post-ejaculatory increases in penile hypersensitivity, continued thrusting for many males can be unpleasant following ejaculation. Post-ejaculatory thrusting can also be diminished as a consequence of an inability to sustain an erection. Many males lose penile tumescence after they ejaculate. The refractory period, as measured by the inability to achieve another erection following ejaculation, also may function to minimize self-semen displacement.

As another adaptation to self-semen displacement, we predicted that males who do not withdraw and continue to thrust past the point of ejaculation would show post-ejaculatory thrusting that was shallower and less vigorous. In contrast to deep thrusting, we found that shallow thrusting with prosthetic genitals failed to produce semen displacement (Gallup *et al.*, 2003). In a recent survey administered to 180 sexually active college students, we discovered that 72% of the males and 87% of the females reported that thrusting became noticeably shallower and less intense following ejaculation (Gallup *et al.*, 2005). Therefore, there is evidence for the existence of a variety of mechanisms in human males that operate to minimize self-semen displacement.

The intra-pair copulation (IPC) proclivity model of female infidelity

One way to think about semen displacement in particular, and sperm competition in general, is in terms of the competing reproductive interests of men and women. If a woman in a committed relationship engages in an extra-pair copulation, not only might this pit the semen of the resident male against the semen of the rival male, it also raises a number of interesting issues relative to the timing and topography of subsequent, and perhaps compensatory, inseminations by the resident male.

The effectiveness of sperm competition and semen displacement as means of competing for paternity is critically related to the elapsed time since the extra-pair encounter. In order to substitute their semen for those of their rivals, we expect resident males to show a high propensity to initiate relatively immediate copulation with their partner under conditions in which they have reason to question her fidelity. On the other hand, if females (consciously or not) engage in extra-pair copulations as a means of using high-quality and/or genetically different males to cuckold their mates, then because of sperm competition and the potential for semen displacement to prevent impregnation, we expect females to have been selected to avoid copulating with their in-pair partners on the heels of an extra-pair copulation. In other words, females should attempt to avoid sex with their committed partners for a period of time following an incidence of infidelity as a means of unwittingly maximizing the likelihood of impregnation by the extra-pair male.

Figure 7.1 represents a depiction of these hypothetical sex differences in the propensity to engage in intra-pair copulations as a function of the elapsed time since the female's sexual encounter with another male. The resident male's insistence on relatively immediate copulation with his partner following a perceived instance of female infidelity is consistent with the importance of minimizing the time between her encounter with the interloper and the application of sperm-competition mechanisms and semen displacement as a means of preventing paternity by his rival.

There is evidence that even sexual intercourse itself may interfere with embryo implantation, which typically occurs within about 24 h of conception.

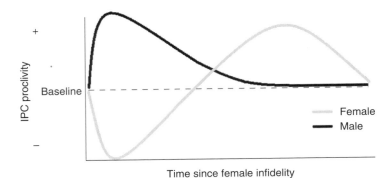

Figure 7.1 The intra-pair copulation (IPC) proclivity model of female infidelity.

It has been known for some time that sexual intercourse, and especially the occurrence of female orgasm, has the effect of increasing uterine myometrial activity (Fox, Wolff, & Baker, 1970). Recent data derived from *in vitro* fertilization embryo-transfer studies indicate that these coitus-induced uterine contractions in human females can interfere with early embryo implantation (Tremellen *et al.*, 2000). Our earlier findings, that males thrust deeper and more vigorously following periods of separation from their partners and following allegations of infidelity (Gallup *et al.*, 2003), may bear on this effect. Not only would deeper and more vigorous penile thrusting effect more complete displacement of rival semen, it may also function to enhance uterine contractions and thereby increase the likelihood of blocking or preventing early embryo implantation. Thus, under conditions of female infidelity, the importance of a short-latency, male-initiated intra-pair copulation may be three-fold. In addition to the operation of (1) sperm competition and (2) semen displacement as a means of minimizing conception by rival males, (3) by promoting uterine contractions, sexual intercourse itself along with the occurrence of female orgasm has the potential to interfere with early implantation of embryos conceived by rival males.

Because of male counter-insemination tactics, we expect females to have been selected to refrain from or at least postpone copulation with the resident male following an extra-pair encounter as a means of enhancing the likelihood of paternity by the rival male. As we have suggested (Gallup & Burch, 2004), if you combine a male's insistence for sex with the female's reluctance for sex, it is a recipe for sexual coercion and rape. Indeed, in support of our analysis, Goetz and Shackelford (see Chapter 5 in this volume) have evidence that the incidence of wife rape is enhanced under conditions in which males believe their wives have been unfaithful.

As a preliminary test of our prediction about females becoming refractory for intra-pair copulations following extra-pair copulations, we recently attempted to measure changes among college females in the propensity to engage in an intra-pair copulation as a function of the time since an instance of infidelity (Gallup et al., 2005). In response to questions about infidelity, 84.2% of the females ($n = 136$) indicated that they would wait at least 48 h or longer before resuming sex with their committed, in-pair partners. Thus, consistent with our model, reluctance to engage in an intra-pair copulation following an extra-pair copulation appears to be a robust and widespread phenomenon among college females which, coupled with the fact that females are more likely to have extra-pair copulations during the ovulatory phase of their cycles and are less likely to use contraceptives (Baker & Bellis, 1995), suggests that women behave (wittingly or not) in ways that enhance the likelihood of conception by extra-pair males.

Another interesting prediction that can be derived from this model concerns the female's recovery and, indeed, eventual resurgence of intra-pair copulation receptivity. Should impregnation and implantation occur as a consequence of an extra-pair encounter, it becomes important for the female to resume copulation with the resident male to mask or obscure the possibility of failed paternity. Thus, the model (see Figure 7.1) not only predicts an eventual return to intra-pair copulation proclivity baseline by both males and females, but as the time since the extra-pair copulation continues to increase, females are predicted to initiate patterns of copulation with the resident male. In other words, if a woman has been impregnated by another man, it becomes important to take steps (consciously or not) to ensure that paternity in the eyes of the resident male is assured. The frequent anecdotal reports that positive feelings toward their committed partners are enhanced as a consequence of partner swapping (Fang, 1976; Wachowiak & Bragg, 1980), may be a byproduct of this compensatory increase in sexual attraction that females experience toward their committed partners on the heels of extra-pair sexual encounters.

Paternal-assurance tactics

To frame the issue of semen displacement as a sperm-competition mechanism in broader perspective, we next comment briefly on the rich ensemble of evolved strategies that may function to assure human paternity. As shown in Table 7.1, there appear to be at least four major categories of paternal-assurance tactic that have emerged during human evolutionary history.

The first and most obvious paternal assurance tactics are those that involve *insemination-prevention strategies*. These include putting a premium on virginity in a bride (for evidence, see Hughes & Gallup, 2003), mate guarding and mate

Table 7.1. *Human paternal-assurance tactics.*

Insemination-prevention strategies
- Preference for virgin brides
- Mate guarding
- Male sexual jealousy
- Mechanical and surgical intervention
- Frequent copulation

Counter-insemination strategies
- Sperm competition
- Semen displacement

Pregnancy-termination strategies
- Coitus-induced uterine contractions
- Pregnancy-induced domestic violence

Postpartum investment strategies
- Paternal resemblance
- Child abuse
- Infanticide

monitoring, mechanical (e.g. chastity belts) and surgical techniques (e.g. infibulation) designed to dissuade female infidelity (Daly, Wilson & Weghorst, 1982), engaging in frequent copulation (Shackelford, 2003), and male sexual jealousy (Daly, Wilson & Weghorst, 1982), all of which function to reduce the likelihood of cuckoldry by minimizing the woman's exposure to semen from rival males. However, if these mechanisms fail and insemination of the female occurs as a consequence of an extra-pair copulation, *counter-insemination strategies*, such as sperm competition (Baker & Bellis, 1995; Birkhead, 2000) and semen displacement (Gallup *et al.*, 2003), come into play. If these counter-insemination tactics fail and impregnation occurs as a consequence of female infidelity, there also appears to be a class of *pregnancy-termination strategies*. These tactics include coitus-induced uterine contractions that interfere with early embryo implantation (Tremellen *et al.*, 2000), and instances of pregnancy-induced domestic violence. There is growing evidence that males who question their partner's fidelity show an increase in spouse abuse during pregnancy, and the abuse is often directed toward the female's abdomen (Burch & Gallup, 2000). Finally, in instances of failed paternity, where children sired by rival males are born, a set of *postpartum investment strategies* come into play. These include the recent discovery that males, but not females, make hypothetical investment decisions that favor children with whom they share facial characteristics (e.g. Platek *et al.*,

2003). Likewise, instances of male-initiated child abuse (e.g. Burch & Gallup, 2000) and infanticide by non-genetic fathers (Daly & Wilson, 1988) serve to reduce and even terminate investment in children of dubious paternity.

The existence of such a wide and diverse array of paternal-assurance tactics suggests strongly that female infidelity was widespread, and as a consequence competition among males for paternity was a prominent, recurring feature of human evolutionary history.

References

Baker, R. R. and Bellis, M. A. (1995). *Human Sperm Competition: Copulation, Masturbation, and Infidelity*. London: Chapman and Hall.

Birkhead, T. (2000). *Promiscuity: an Evolutionary History of Sperm Competition*. Cambridge, MA: Harvard University Press.

Burch, R. L. and Gallup, G. G., Jr. (2000). Perceptions of paternal resemblance predict family violence. *Evolution and Human Behavior*, **21**(6), 429–35.

Daly, M. and Wilson, M. (1988). Evolutionary social psychology and family homicide. *Science*, **242**, 519–24.

Daly, M., Wilson, M. and Weghorst, S. (1982). Male sexual jealousy. *Ethology and Sociobiology*, **3**, 11–27.

Eberhard, W. G. (1985). *Sexual Selection and Animal Genitalia*. New Haven, CT: Harvard University Press.

Fang, B. (1976) Swinging: in retrospect. *The Journal of Sex Research*, **12**, 220–37.

Ford, C. S. and Beach, F. (1951). *Patterns of Sexual Behavior*. New York: Harper and Row.

Fox, C. A., Wolff, H. S., and Baker, J. A. (1970). Measurement of intra-vaginal and intra-uterine pressures during human coitus by radio-telemetry. *Journal of Reproduction and Fertility*, **22**(2), 243–51.

Gallup, G. G., Jr. and Burch, R. L. (2004). Semen displacement as a sperm competition strategy. *Evolutionary Psychology*, **2**, 12–23.

Gallup, G. G., Jr. and Suarez, S. D. (1983). Optimal reproductive strategies for bipedalism. *Journal of Human Evolution*, **12**, 193–6.

Gallup, G. G., Jr., Burch, R. L., Zappieri, M. L., *et al.* (2003). The human penis as a semen displacement device. *Evolution and Human Behavior*, **24**, 277–89.

Gallup, G. G., Jr., Burch, R. L., and Beren Mitchell, T. (2005). Semen displacement as a sperm competition strategy: multiple mating, self-semen displacement, and timing of in-pair copulations. *Human Nature* (in press).

Gebhard, P. H. and Johnson, A. B. (1979). *The Kinsey Data: Marginal Tabulations of the 1938–1963 Interviews Conducted by the Institute for Sex Research*. Philadelphia, PA: W. B. Saunders.

Gonzales G. F., Kortebani G., and Mazzolli A. B. (1993). Hyperviscosity and hypofunction of the seminal vesicles. *Archives of Andrology*, **30**, 63–8.

Gopalkrishnan, K., Padwal, V., and Balaiah, D. (2000). Does seminal fluid viscosity influence sperm chromatin integrity? *Archives of Andrology*, **45**, 99–103.

Hughes, S. M. and Gallup, G. G., Jr. (2003). Sex differences in morphological predictors of sexual behavior: shoulder to hip and waist to hip ratios. *Evolution and Human Behavior*, **24**, 173–8.

Hughes, S. M., Harrison, M. A., and Gallup, G. G., Jr. (2004). Sex differences in mating strategies: mate guarding, infidelity, and multiple concurrent partners. *Sexualities, Evolution, & Gender*, **6**, 3–13.

McFarlane, J. and Wilson, P. (2000). Intimate partner violence. *Journal of Interpersonal Violence*, **15**(2), 158–70.

Platek, S. M., Critton, S. R., Burch, R. L., *et al.* (2003). How much resemblance is enough? Sex difference in reactions to resemblance, but not the ability to detect resemblance. *Evolution and Human Behavior*, **24**, 81–7.

Shackelford, T. K. (2003). Preventing, correcting, and anticipating female infidelity: three adaptive problems of sperm competition. *Evolution and Cognition*, **9**, 90–6.

Stockwell, M., Platek, S. M., Burch, R. L., and Gallup G. G., Jr. (2000). *Variation in male sexual jealousy as a function of race.* Poster presented at the Human Behavior and Evolution Society Annual Meeting, Amherst, MA.

Stockwell, M., Burch, R. L., Platek, S. M., and Gallup G. G., Jr. (2001). *Racial differences in postcopulatory behavior.* Poster presented at Eastern Psychological Association, Washington, DC.

Straus, M. and Gelles, R. (1986). Societal change and change in family violence from 1975 to 1985 as revealed by two national surveys. *Journal of Marriage and the Family*, **48**, 465–79.

Tremellen, K. P., Valbuena, D., Landeras, J., *et al.* (2000). The effect of intercourse on pregnancy rates during assisted human reproduction. *Human Reproduction*, **15**(12), 2653–8.

Wachowiak, D. and Bragg, H. (1980). Open marriage and marital adjustment. *Journal of Marriage and the Family*, **42**, 57–62.

Wilson, M. and Daly, M. (1992). The man who mistook his wife for a chattel. In J. H. Barkow, L. Cosmides, and J. Tooby, eds, *The Adapted Mind: Evolutionary Psychology and the Generation of Culture.* New York, NY: Oxford University Press, pp. 289–326.

8

The psychobiology of human semen

REBECCA L. BURCH
SUNY at Oswego
AND
GORDON G. GALLUP, JR.
SUNY at Albany

Introduction

Our interest in the psychological properties of semen arose as a byproduct of an initial interest in menstrual synchrony. In reviewing that literature we discovered several articles (Trevathan, Burleson, & Gregory, 1993; Weller & Weller, 1998) reporting that lesbians who live together fail to show menstrual synchrony. Since the evidence suggests that menstrual synchrony is mediated by the exchange of subtle olfactory cues among cohabitating women (Preti *et al.*, 1986, Stern & McClintock, 1998) this struck us as peculiar, because lesbians would be expected to be in closer, more intimate contact with one another on a daily basis than other females who live together. What is it about heterosexual females that promotes menstrual synchrony, or conversely what is it about lesbians that prevents menstrual synchrony? It occurred to us that one feature that distinguishes heterosexual women from lesbians is the presence or absence of semen in the female reproductive tract. Lesbians have semen-free sex.

Human semen is a very complicated mixture of many different ingredients. If you extract the sperm from semen, what is left is called seminal plasma. We speculated that there may be chemicals in seminal plasma that, through vaginal absorption, affect female biology and triggers the release of pheromones that function to entrain menstrual cycles among cohabitating women. Some of the components in semen pass through vaginal epithelial tissue, and within an hour or two after intercourse heightened levels of certain seminal chemicals can be detected in the female bloodstream (Benziger & Edelson, 1983). Thus, among a pair of cohabitating females the one with semen in her reproductive tract may entrain/drive her roommate's menstrual cycle. This might also explain the variance in menstrual synchrony among women, with some females who show it and some who do not as a function of whether they are sexually active and whether they are using condoms (Gallup, Burch & Platek, 2002). Other studies have linked heterosexual sexual activity with menstrual cycle length (Cutler, Garcia, & Krieger, 1979) and menstrual synchrony (Jarett, 1984; Matteo, 1987).

In reviewing the literature on semen chemistry we discovered an article by Ney (1986) who, on the basis of a depressed female patient he was seeing, speculated that human semen may have antidepressant properties. It occurred to us that one way to test this hypothesis would be to compare levels of depression in women as a function of whether they were using condoms (Gallup *et al.*, 2002). Sexually active heterosexual females who use condoms and lesbians share one important feature in common, they are both having semen-free sex. Using scores on the Beck Depression Inventory to index depression, we discovered that sexually active female college students who were not using condoms were significantly less depressed than those who were. Interestingly, those who were using condoms did not differ in terms of their BDI scores from those who were not having sex. Consistent with the possibility that semen may have antidepressant properties, BDI scores among those who were not using condoms were correlated with the elapsed time since their last sexual encounter (i.e. as the time since their last exposure increased, so did their depressive symptoms). When we took into account those who were using hormonal contraceptives and those who were in committed relationships, the differences in depression among those who were not using condoms and those who were did not prevail.

Semen chemistry: a brief overview

It is possible that the presence of semen in the reproductive tract of women not only affects depressive symptoms, as Ney suggested, but may trigger physiological changes in menstrual cycle length, variability, and synchrony, perimenopausal symptoms, and psychological changes in menopausal and postpartum depression, premenstrual syndrome, and female-initiated sexual activity. The following compounds have been identified as being present in human semen. Each of these compounds has also been shown to affect female sexual behavior and physiology. A list of the compounds and their concentrations are shown in Table 8.1.

First of all cholesterol, the precursor to all steroid hormones, is found in semen (Valsa, Skandhan, & Umarvanshi, 1992). Concentrations of cortisol (formed from cholesterol) in human seminal plasma, as estimated by immunoassay, were about 60% of random levels in blood serum (Brotherton, 1990a).

Table 8.1. *Annotated list of compounds and their concentrations in human seminal plasma.*

Compound	Level (ng/l)	Reference	Function
Cholesterol	6148	Valsa et al. (1992)	Steroid hormone precursor.
Cortisol	63 700	Brotherton (1990a)	Increase approach behaviors, parental care, affection.
Cortisol	20 000	Abbaticchio et al. (1981)	
Transcortin	12 000 000	Brotherton (1990a)	Stress response, accentuate dopamine effects, increases oxytocin, corticotropin-releasing hormone, and opioids.
Testosterone	500	Hampl et al. (2000)	Sexual drive.
Testosterone	559	Asch et al. (1984)	
Dihydrotestosterone	695	Asch et al. (1984)	
5α-Dihydrotestosterone	300	Hampl et al. (2000)	
Androstenedione	2018	Kwan et al. (1992)	Precursor to other steroids, immunological properties.
5α-Androst-16-en-3α-ol	600	Kwan et al. (1992)	Precursor to other steroids, immunological properties.
5α-Androst-16-en-3β-ol	600	Kwan et al. (1992)	Precursor to other steroids, immunological properties.
5,16-Androstadien-3β-ol	600	Kwan et al. (1992)	Precursor to other steroids, immunological properties.
5α-Androst-16-en-3-one	800	Kwan et al. (1992)	Precursor to other steroids, immunological properties.
4,16-Androstadien-3-one	800	Kwan et al. (1992)	Precursor to other steroids, immunological properties.
7α-Dihydroxy-5-androsten-17-one	1808.176	Hampl et al. (2000)	Precursor to other steroids, immunological properties.
7β-Dihydroxy-5-androsten-17-one	1694.969	Hampl et al. (2000)	Precursor to other steroids, immunological properties.
Estrone	157	Ney (1986)	Trigger ovulation, elevate mood, vaginal lubrication, absorption of hormones, pheromone production.
Estradiol	70.2	Ney (1986)	Steroidogenesis, egg development, ovulation.
Estradiol	46.9–91.3	Luboshitzky et al. (2002)	
Follicle-stimulating hormone	8 500 000	Ney (1986)	Steroidogenesis, egg development, ovulation.
Luteinizing hormone	220 000 000	Ney (1986)	Steroidogenesis, egg development, ovulation, sex drive.

Table 8.1. (cont.)

Substance	Concentration	Reference	Function
Luteinizing hormone-releasing hormone	31–71	Chan & Tang (1983)	Induces luteinizing hormone release.
Prolactin	86 000	Ney (1986)	Antidepressant, facilitates pregnancy, maternal behavior, appetite, oxytocin secretion, and ACTH secretion.
Prolactin	7500	Aiman et al. (1988)	Immunosuppressant.
19-Hydroxyprostaglandin E		Kelly (1995)	
Prostaglandin E_1		Ney (1986), Kelly (1995)	Uterine contractions, ovulation.
Prostaglandin E_2		Ney (1986), Kelly (1995)	Uterine contractions, ovulation.
Prostaglandin F_2		Ney (1986)	Uterine contractions, ovulation.
Oxytocin	1.72	Goverde et al. (1998)	Increases testosterone, prostaglandins, involved in orgasm, affiliation and bonding, elevates mood.
Vasopressin	1.84	Brotherton (1990c)	Arousal, attention, vigilance, sympathetic function.
Thyrotropin-releasing hormone	12 200	Pekary et al. (1983)	Antidepressant.
Melatonin	9.7–45.4	Luboshitzky et al. (2002)	Increases steroid effects, affects reproduction, induces sleep.
Relaxin	1240–73 000	Loumaye et al. (1980)	Sperm motility, fertilization, implantation, uterine growth and accommodation.
Human chorionic gonadotropin	232.067–2510.548	de Medeiros et al. (1992)	Pregnancy maintenance.
Human placental lactogen		Seppälä et al. (1985)	Pregnancy maintenance.
Pregnancy-specific β1-glycoprotein		Seppälä et al. (1985)	Pregnancy maintenance.
Placental protein 5		Seppälä et al. (1985)	Pregnancy maintenance.
Placental protein 12		Seppälä et al. (1985)	Pregnancy maintenance.
Placental protein 14		Seppälä et al. (1985)	Pregnancy maintenance.
Pregnancy-associated plasma protein		Seppälä et al. (1985)	Pregnancy maintenance.
Serotonin		Gonzales & Garcia-Hjarles (1990)	Antidepressant.
Tyrosine		van Overveld et al. (2000)	Precursor to dopamine, norepinephrine, epinephrine.

Compound	Value	Reference	Notes
DOPA	40 230	Fait et al. (2001)	Precursor to dopamine.
Norepinephrine	151 810	Fait et al. (2001)	Attention, cognition, concentration.
β-Endorphin	154.7	Zalata et al. (1995)	Immunosuppressant, opioid.
β-Endorphin	308	Davidson et al. (1989)	
β-Endorphin	192	Singer et al. (1989)	
Calcitonin	331	Davidson et al. (1989)	
Calcitonin	754	Singer et al. (1989)	
Enkephalin		Sastry et al. (1991)	Prevents acrosome activation, opioid, orgasm.
Substance P		Sastry et al. (1991)	
Cytokine interleukin 1α	26.1	Maegawa et al. (2002)	Immunosuppressant.
Cytokine interleukin 2		Maegawa et al. (2002)	Immunosuppressant.
Cytokine interleukin 4		Maegawa et al. (2002)	Immunosuppressant.
Cytokine interleukin 6		Maegawa et al. (2002)	Immunosuppressant.
Cytokine interleukin 8		Maegawa et al. (2002)	Immunosuppressant.
Tumor necrosis factor-α		Maegawa et al. (2002)	Immunosuppressant.
Interferon-γ		Maegawa et al. (2002)	Immunosuppressant.
Granulocyte colony-stimulating factor		Maegawa et al. (2002)	Immunosuppressant.
Macrophage colony-stimulating factor		Maegawa et al. (2002)	Immunosuppressant.
Granulocyte elastase		Maegawa et al. (2002)	Immunosuppressant.
Carnitine		Ruiz-Pesini et al. (2001)	
Carnitine	256.918	Menchini-Fabris et al. (1984)	

Abbaticchio *et al.* (1981) also found that levels of cortisol in the blood prove to be much greater than in the seminal plasma. The researchers found no significant differences between normozoospermic and oligo-azoospermic subjects, either in the blood, or in the seminal plasma. In other words, both fertile men and men with fertility issues have similar levels of cortisol. Transcortin (corticosteroid-binding globulin) has also been found in human seminal plasma, although transcortin concentrations were only about 10% of levels in blood serum (Brotherton, 1990a).

Corticoids are best known for their role in response to stress. The adrenal cortex releases glucocorticoids (as well as prolactin, thyroid hormones, and vasopressin) in response to virtually any stressor (Nelson, 2000). Glucocorticoids accentuate the effects of dopamine, increasing reinforcement for behaviors. Blood levels of cortisol have been shown to increase approach behaviors, including parental care and interpersonal affection. This is thought to be due to an increased arousal and reinforcement of the stimuli the person is experiencing. Although there have been examples of glucocorticoid involvement in the induction of parental care, it is possible that the changes are due to the increases in oxytocin, corticotropin-releasing hormone, and opioids which accompany glucocorticoid release (Nelson, 2000).

Testosterone is in relatively high concentration in human semen compared to other compounds, approximately 559 pg/ml (Ney, 1986). Other androgens are also present, some in higher concentrations (see Table 8.1). These compounds can be used to derive other steroids and they themselves appear to have immunological properties. Levels of testosterone in semen are also correlated with sperm motility. In fact, seminal concentrations of testosterone and dihydrotestosterone were significantly higher in subjects with sperm in their ejaculate than in vasectomized men (Asch *et al.*, 1984). Purvis *et al.* (1975) found a positive correlation between testosterone and dihydrotestosterone levels of the seminal plasma of normal and azoospermic subjects, indicating that one may increase the other. Another interesting finding regarding androgens in seminal plasma indicates that the ratio between testosterone and dihydrotestosterone is different from one person to the other, due to the heterogeneity of seminal plasma which stems for the most part from the male accessory sex glands, and the prostate and seminal vesicles. This ratio may be useful in identifying the person's semen (Doss & Louca, 1991).

Testosterone is absorbed by the vagina, and at higher levels than testosterone administered transdermally. Approximately 63% of the testosterone administered vaginally was absorbed (Wester, Noonan, & Maibach, 1980). There are suggestions in the literature that normal variations in androgens during the menstrual cycle could be responsible for reported cyclic fluctuations in sexual interest (Guay, 2001; Van Goozen *et al.*, 1997). The adrenal cortex produces significant quantities of androgens, and adrenalectomy is detrimental to female sexual behavior (Dixson, 1987; Hepburn *et al.*, 1996). Furthermore, since the 1940s exogenous androgen treatments have been given to human females for diverse medical purposes. Even in very low doses, androgen treatments can increase sexual motivation in some human females (Carter, 1992). Morris *et al.* (1987) measured testosterone levels in 43 married women. Testosterone levels correlated significantly with sexual intercourse frequency.

Human semen contains both estrone and estradiol (Asch *et al.*, 1984), although the proportion of estrone is more than twice that of estradiol (Ney, 1986). According to Luboshitzky, Shen-Orr, and Herer (2002), estradiol was in significantly higher concentration in seminal than blood plasma in males. When estrone was administered in the vagina, it was rapidly absorbed and plasma levels increased 24-fold. These high levels were maintained for 2 h (Schiff, Tulchinsky & Ryan, 1977). Rigg *et al.* (1977) administered 17-β-estradiol and observed a rapid peak concentration in the blood 110 times the basal level. This also lasted for 2 h. Research also suggests that the vaginal administration of estrogen assists in the absorption of other compounds, including progesterone (Villanueva, Casper, & Yen, 1981).

Estrogen, along with luteinizing hormone and follicle-stimulating hormone, work in concert to trigger ovulation in females and, along with luteinizing hormone and follicle-stimulating hormone, reaches its peak at ovulation. Peak fertility more or less coincides with the estrogen peak (Nelson, 2000). Estrogen, whether secreted by the ovarian follicle or given as replacement therapy, may have broad effects on behavior. Estrogen apparently can increase female sexual motivation and may facilitate peripheral changes such as the production of odors that make the female more attractive to the male partner. In human females, estrogen also increases vaginal lubrication and thus can influence sexual behavior indirectly (Carter, 1992).

Deficits in estrogen have been associated with depression. In one study, over 90% of the depressed women treated with estrogen significantly improved their mood (Klaiber *et al.*, 1979). It is important to note that high, pharmacological doses were used in this study. Administration of estrogen in physiological doses improves mood in normal women (Sherwin & Gelfand, 1985) but not in clinically depressed women (Schneider *et al.*, 1997). A great deal of research has found that estrogen is effective in the treatment of menopausal depression (Lebowitz, Pollock & Schneider, 1997; Lebowitz *et al.*, 1997). In a meta-analysis conducted by Zweifel and O'Brien (1997), the effect of exogenous estrogen and other steroidal hormones were related to the alleviation of depressed mood. Progesterone alone, and in combination with estrogen, was associated with smaller reductions in depressed mood and androgen alone or in combination with estrogen was associated with greater reductions in depressed mood.

The most interesting facet of luteinizing hormone in semen is its astounding concentration. Semen contains 220 mIU/ml luteinizing hormone, the highest concentration of any hormone. Follicle-stimulating hormone is found in a much lesser concentration (Ney, 1986). Researchers have found that concentrations of luteinizing hormone in seminal plasma are as much as five times higher than in the blood (Mondina *et al.*, 1976; Sheth, Shah, & Mugatwala, 1976). This is rather unusual, since most other peptide hormones in semen are in equal or lower concentrations than in the blood. Luteinizing hormone in semen is also linked to higher numbers and motility of sperm, compounding the male's fertility (Asch *et al.*, 1984). Human seminal plasma also contains luteinizing hormone-releasing hormone (Chan & Tang, 1983).

Little research has been done to determine levels of vaginal absorption of these hormones. The few studies that do mention follicle stimulating hormone and luteinizing hormone only discuss the resultant decreases in levels after the administration of high, pharmacological doses of estrogens (Keller *et al.*, 1981). Follicle-stimulating hormone and luteinizing hormone (both glycoproteins) are structurally similar and both stimulate steroidogenesis in the gonads, as well as the development and maturation of gametes. In other words, they both act in females to produce and release eggs, and both peak at ovulation (Nelson, 2000). The deposition of follicle-stimulating hormone and luteinizing hormone, and their subsequent absorption into the female bloodstream could act to facilitate or even induce ovulation (see below).

Little research has also been conducted on the vaginal absorption of prolactin. However, the absorption and subsequent rise in estrogen levels triggers an increase in prolactin as well (Keller *et al.*, 1981; Yamazaki, 1984). Human semen contains approximately 86 ng/ml prolactin (Ney, 1986). Seminal plasma prolactin concentrations are related directly to sperm concentrations and motilities and are much lower in men suffering from infertility (Aiman, McAsey, & Harms, 1988). However, other researchers (Asch *et al.*, 1984) found no differences between fertile and vasectomized men in seminal concentrations of prolactin.

Prolactin is a hormone common throughout vertebrate evolution, and can have hundreds of unique physiological functions including effects during pregnancy and lactation (Nelson, 2000). Prolactin has been reported to influence numerous brain functions, including maternal behavior, feeding and appetite, oxytocin secretion, and ACTH (corticotropin) secretion in response to stress. Hence, prolactin may be a key player in the coordination of neuroendocrine and behavioral adaptations of the maternal brain (Grattan, 2001).

Prolactin, as studied in rodents, aids in the formation of the corpora lutea and also seems to potentiate the release of progesterone from the corpus luteum, which facilitates pregnancy. Deficiencies in prolactin release involve the mono-amine neurotransmitter systems that have been implicated in depression (Golden *et al.*, 2002; Nelson, 2000). Deficiencies in prolactin secretion are also implicated in postpartum depression (Hendrick, Altshuler & Suri, 1998) and premenstrual syndrome (Derzko, 1990).

Although Homberg and Samuelsson (1966) isolated the presence of 13 differ-ent prostaglandins in human semen, little research has focused on these sub-stances. However, some prostaglandins have been shown to be absorbed rapidly through the vagina, namely E_1, E_2, and F_{2a} (Sandberg *et al.*, 1968). Increased frequency of uterine contraction has resulted from prostaglandin E_1 adminis-tration. Certain prostaglandins, F_{1a} and F_{2a}, are active in the dissolution of the corpora lutea and in ovulation (Nelson, 2000) and these compounds cause powerful contractions of the uterus (Mackenzie, Bradley, & Mitchell, 1980). Based on this information, it is possible that prostaglandins in semen are used to assist in ovulation. Ney (1986) also hypothesized that prostaglandins assisted in alleviating depression, citing Abdullah and Hamadah (1975) who found that drug-free patients with depression possessed significantly less prostaglandin E_1, while manic patients possessed significantly more.

Other studies have shown prostaglandins to have immunosuppressive cap-abilities (Kelly, 1995). Skibinski *et al.* (1992) found that prostaglandins E_1 and E_2 exerted a greater immunosuppressive effect than 19-OH prostaglandin E, but considerably higher levels of 19-OH prostaglandin E in semen might contribute to the majority of immunosuppressive activity *in vivo*. Thus, with the presence of prostaglandins, immunological reactions would be inhibited for a period of time after intercourse.

Other suppressive agents are present in semen and may exert specific effects (Kelly, 1995). Maegawa *et al.* (2002) reported a repertoire of cytokines in seminal plasma (see Table 8.1). Each of these immunosuppressants may act in the female to dull her immunological reaction to the foreign bodies (sperm) invading her vagina and cervix. Thus the chemistry of human semen would appear to include components that increase the likelihood of impregnation.

Human semen contains large amounts of opioid peptides (such as β-endor-phin) and cytokines. Zalata *et al.* (1995) concluded that β-endorphin in seminal plasma may play an immune-suppressive role. Singer *et al.* (1989) suggested that β-endorphin and calcitonin may affect sperm motility. Mungan *et al.* (2001) also studied levels of calcitonin in seminal fluid and its effects on sperm motility and found that seminal calcitonin levels were significantly correlated with sperm motility. While these endorphins may assist in sperm motility, it is important to investigate the role that these chemicals play in the female immunological response and any possible analgesic properties the compounds may have when absorbed through the vagina.

Enkephalins are one of the opioids present in human semen and to date their function in this context remains unknown (Fernandez *et al.*, 2002). Sastry, Janson, and Owens (1991) found high levels of leucine enkephalin in human seminal plasma as well as lower levels of substance P. While substance P increased sperm motility, leucine enkephalin depressed it. It was concluded that substance P-like tachykinins may play a role in sperm maturation, in expulsion of fluid from the epididymis, and in initiation of motility, whereas leucine enkephalin-like peptides may contribute to the orgasmic experience and detumescence. It is not currently known whether these peptides would affect the orgasmic experience or postcoital experiences of the female. Endorphins and enkephalins have numerous physiological effects, but primarily they decrease anxiety, induce analgesia and drowsiness, and assist in immune function and reinforcing effects (Stahl, 2001). Absorption rates and levels in the female bloodstream during and after coitus still need to be examined.

Oxytocin is found in the seminal plasma of normal men and is slightly lower in patients with poor semen quality and vasectomized patients. No statistically significant relationships have been found between the oxytocin levels and sperm characteristics (Goverde *et al.*, 1998). Oxytocin has been shown to have wide-ranging effects in humans (Nelson, 2000). Although it is well known for its effects during parturition and lactation, oxytocin also increases production of other hormones, such as prostaglandins and testosterone, and assists in the stimulation of ovulation and blastocyst development. Oxytocin has been thought of as an "affiliation hormone" because research on non-human mammals has demonstrated that it plays a key role in the initiation of maternal behavior and the formation of adult pair bonds. Oxytocin is implicated in male penile erection and female orgasm as well as the social aspects of romantic relationships (Turner *et al.*, 1999). It is possible that oxytocin, transferred to the female during coitus and absorbed, can strengthen the pair bond and make the sexual activity more rewarding.

Vasopressin, a peptide hormone similar in structure to oxytocin, has also been identified in semen. In the animal literature, vasopressin has been associated with behaviors that might be classified broadly as "defensive," including enhanced arousal, attention, or vigilance, increased aggressive behavior, and a general increase in sympathetic functions (Carter & Altemus, 1997; Nelson, 2000). Virtually nothing is known regarding the effects in humans of centrally administered vasopressin because it does not readily pass through the blood–brain barrier (Carter & Altemus, 1997). In animals, vasopressin has been implicated in the central mediation of complex social behaviors, including affiliation, parental care, and territorial aggression. Intense aggression towards strangers for defense of territory, nest, and mate has long been an identifying feature of monogamy in these species (Winslow et al., 1993). It is possible that vasopressin, like oxytocin, can act to strengthen pair bonds, and may induce parental behaviors and sexual jealousy.

A variety of placental proteins, including human chorionic gonadotropin (Asch et al., 1984; de Medeiros et al., 1992), human placental lactogen, pregnancy-specific β1-glycoprotein, placental protein 5 (Seppälä et al., 1985), and ferritin (Brotherton, 1990b), have been found in human seminal plasma. In many cases their concentrations in follicular fluid and seminal plasma greatly exceeded those in the serum of nonpregnant women or men, and sometimes they even exceeded pregnancy levels. Concentrations of β-human chorionic gonadotropin and luteinizing hormone were highly correlated with the numbers and motility of sperm in the ejaculate (de Medeiros et al., 1992). It is possible that these levels of human chorionic gonadotropin can increase the probability of conception and pregnancy maintenance in women. Seppälä et al. (1985) also found a number of pregnancy-associated proteins (see Table 8.1) in follicular fluid and seminal plasma. The levels of placental protein 5 in seminal plasma showed an association with sperm motility, suggesting that placental protein 5 may have a significant biological function in the maintenance of sperm motility (Lee et al., 1983).

Relaxin is a polypeptide hormone produced in the human female by the corpus luteum and deciduas during pregnancy. In the male it is produced in the prostate and is present in human semen (MacLennan, 1991). According to Stewart et al. (1990), relaxin is significantly elevated 9–10 days following ovulation in women. Relaxin also increases 1–2 days prior to the first detectable increase in plasma human chorionic gonadotropin during pregnancy. Relaxin may have significant roles in sperm motility, fertilization, implantation, uterine growth and accommodation, the control of myometrial activity to prevent preterm labor, cervical ripening, and the facilitation of labor (MacLennan, 1991; Weiss, 1995). Its role in pregnancy maintenance has yet to be elucidated fully, but relaxin in seminal plasma may act to manipulate the female reproductive cycle or facilitate pregnancy. It is interesting that compounds defined by their pregnancy-maintaining effects are present in semen, implying that repeated insemination after conception may play a role in producing positive pregnancy outcomes (see Chapter 10 in this volume).

Pekary, Hershman, and Friedman (1983) reported the presence of high levels of thyrotropin-releasing hormone and a thyrotropin-releasing-hormone-homologous peptide, in the human prostate. Others have reported a thyrotropin-releasing-hormone-like peptide in human semen (Gkonos *et al.*, 1994; Khan *et al.*, 1992; Khan & Smyth, 1993; Pekary *et al.*, 1990). Gkonos *et al.* (1994) suggest that this thyrotropin-releasing-hormone-like peptide may play a role in human reproductive physiology. In addition, thyroid hormones, and specifically thyrotropin-releasing hormone, have been utilized in the investigation and treatment of depression. Administration of thyrotropin-releasing hormone stimulates the release of thyroid-stimulating hormone from the anterior pituitary gland and subsequent hormone production by the thyroid gland. In fact, administration of thyrotropin-releasing hormone has also been shown to successfully treat some sufferers of premenstrual syndrome (Roy-Byrne *et al.*, 1984). Depressed patients have a significantly lower thyroid response to thyroid-stimulating hormone and show high levels of antibodies against thyrotropin-releasing hormone. Depressive symptoms are also treated with thyrotropin-releasing hormone (Nelson, 2000).

Serotonin has been found in human seminal plasma, and increases sperm motility, but extremely high levels have been correlated with certain types of infertility (Gonzales *et al.*, 1989; Gonzales & Garcia-Hjarles, 1990). However, seminal volume, pH, sperm morphology, fructose, citric acid, and serum testoterone values were similar between groups of patients with different levels of blood serotonin. This suggests an optimal level of serotonin in semen for male reproductive physiology.

Although the role of serotonin in depression, and the treatment of depression, is well documented (Stahl, 2001), little to no information is available on activity of seminal serotonin in the female body, absorption levels, metabolism, or physiological reactions. Even though serotonin does not pass through the blood–brain barrier, it may affect peripheral sites and act indirectly to alter emotions and behavior. Since depression can be characterized as a serotonin-deficiency syndrome (Stahl, 2001), the absorption of seminal serotonin could have an effect, at the very least by contributing the metabolized building blocks of serotonin and increasing serotonin synthesis.

Serotonin metabolite melatonin is also found in semen (Yie *et al.*, 1991). Luboshitzky *et al.* (2002) found that melatonin levels in the semen of normal men averaged 0.6–5.0 pg/ml, significantly lower than blood levels. High levels of melatonin have been found in men with oligospermia and azoospermia, which suggests that melatonin may have an effect upon both sperm production and motility, but its effect in normal males remains unclear (Yie *et al.*, 1991).

Several studies have linked melatonin to reproductive behaviors in humans (Nelson, 2000), possibly reflecting seasonal fluctuations in sexual behaviors and other hormonal levels (Cassone *et al.*, 1993). It is also important to the development of the reproductive system (Davis, 1997). It appears that melatonin inhibits the negative-feedback loop of steroid hormones, increasing their levels and effects (Davis, 1997). In fact, melatonin has been found to both stimulate and inhibit reproductive function, depending on how it is released (Weaver, 1997).

Because of its relationship to serotonin, melatonin has also been linked to mood; in particular, it is negatively correlated with incidence of suicide (Souêtre *et al.*, 1987) and deficiencies implicated in Seasonal Affective Disorder. Other research shows that bipolar patients have a more vulnerable and deficient melatonin system (Lewy *et al.*, 1984). There is a strong correlation between clinical symptoms of depressed mood, reality disturbance, and low levels of melatonin (Brown *et al.*, 1987).

By and large, the effects of melatonin are to induce sleepiness and fatigue (Weaver, 1997). It is possible that the addition of seminal melatonin could raise blood levels and function to induce sleep in the female after copulation. This is consistent with evidence showing that the resumption of an upright posture following insemination endangers sperm retention (see Chapter 7 in this volume). Melatonin is lipid-soluble and may be able to act fairly quickly.

Fait *et al.* (2001) found that in addition to the precursor tyrosine, epinephrine, norepinephrine, 3,4-dihydroxyphenylalanine (DOPA), and 3,4-dihydroxyphenyl acetic acid (DOPAC) were found in the semen of healthy volunteers. Norepinephrine and DOPA were present in all specimens. These concentrations are respectively 19 times and twice as high as the normal concentration in plasma. No correlation was found between the concentration of any of the catecholamines and semen characteristics.

These neurotransmitters, along with their precursor tyrosine, have a huge impact on human behavior. A deficiency in such neurotransmitters is the major hypothesis behind depressive disorders. Tyrosine, once absorbed by the body, can easily be metabolized into dopamine, the neurotransmitter released in all rewarding behaviors and responsible for reward and addiction, and norepinephrine, which is involved in greater concentration, attention, and arousal (Stahl, 2001).

Other possible effects of semen

Given the numerous compounds in semen and the documented effects that some of these have on human behavior, it is possible to generate any number of intriguing predictions about the effects of unprotected sexual intercourse in heterosexual women.

FEMALE-INITIATED SEXUAL ACTIVITY

Several seminal compounds have been implicated in affecting libido; testosterone, estrogen, dopamine, opioids, and, in particular, luteinizing hormone. Female-initiated sexual activity seems to peak just before ovulation, during the preovulatory luteinizing hormone surge (Dennerstein et al., 1994; Jarvis & McCabe, 1991; Matteo & Rissman, 1984). The higher concentrations of luteinizing hormone (and luteinizing hormone-releasing hormone) in semen could elevate luteinizing hormone in females after intercourse, resulting in higher levels of female-initiated sexual activity. Testosterone is also correlated with libido in women, and the levels of testosterone in semen could also contribute to this effect (Carter, 1992; Morris et al., 1987). Gallup et al. (2002) found that the women who did not use condoms actually had intercourse most often. Women who used condoms infrequently or not at all not only had more frequent intercourse, but also became more depressed the longer they abstained from intercourse. This was not the result of being in a more committed romantic relationship; both single women and women in relationships showed this pattern, as long as they were not using condoms.

DEPRESSION

The antidepressant effects of semen may be related to estrogen, serotonin, thyrotropin-releasing hormone, the catecholamine neurotransmitters, or even endorphins. Gallup et al. (2002) reported that females who engaged in sexual intercourse but did not use condoms showed significantly lower depression scores on the Beck Depression Inventory than those who used condoms or those who never engaged in sexual intercourse. However, depression scores between females who used condoms and those who did not engage in sexual intercourse were not significantly different. Other variables such as being in a relationship, length of that relationship, use of oral contraceptives, and frequency of sex did not affect depression scores.

POSTPARTUM DEPRESSION

Postpartum depression refers to a nonpsychotic depressive episode that begins in or extends into the period following childbirth (Josefsson et al., 2001) and is probably the most common complication of the puerperium (Lawrie, Herxheimer, & Dalton, 2000). A deficiency or imbalance of sex hormones has repeatedly been suggested as a cause of postpartum depression (Lawrie et al., 2000). Lawrie et al. (2000) found that estrogen therapy in severely depressed women was associated with a greater improvement than placebo. Late pregnancy and the early postpartum period are usually times of abstinence in women. von Sydow (1999) found that, on average, coital activity increased slightly in the first trimester of pregnancy, showed variable patterns in the second trimester, and decreased sharply in the third trimester. After the birth of a child, sexual interest and activity was reduced for several months as compared with the pre-pregnancy level, and sexual problems occurred relatively often.

MENOPAUSAL DEPRESSION

Declining estrogen at menopause, as well as decreasing levels of sexual activity caused by menopausal symptoms such as vaginal atrophy and dryness, may be the primary causes of menopausal depression. Vaginal atrophy and dryness may also affect vaginal absorption of seminal products when sexual activity occurs. Estrogen has been shown to have mood-elevating effects in menopausal women (Burt, Altschuler, & Rasgon, 1998; Coope, 1996; Meyers & Moline, 1997). Sexual activity during menopause decreased vaginal atrophy (Bachmann *et al.*, 1984). Estrogen has also been found to decrease perimenopausal symptoms such as hot flashes (Coope, Thomson, & Poller, 1975; Utian, 1972) and McCoy, Cutler, and Davidson, (1985) found that sexual activity was correlated with decreased hot flashes as well.

Because menopause is a period of reduced sexual activity for many women, it could, like the postpartum period, be thought of as a period of semen withdrawal. Therefore we would predict that women who had unprotected sex would be at greater risk of both menopausal and postpartum depression.

MENSTRUAL CYCLE FLUCTUATIONS IN MOOD AND PREMENSTRUAL SYNDROME

There is a vast amount of literature examining fluctuations in mood, with most finding exacerbation of negative mood during the luteal and menstrual phases (Allen *et al.*, 1996; Bloch, Schmidt, & Rubinow, 1997; Henderson & Whissell, 1997; Williams & Krahenbuhl, 1997). It has been shown that as many as 95% of women experienced depression during the days preceding or at the onset of menstruation (Golub, 1976). Anxious symptoms (Cameron *et al.*, 1986; Cook *et al.*, 1990), including panic attacks (Cameron, *et al.*, 1988) and obsessive/compulsive symptoms (Williams & Koran, 1997) are also more prevalent toward the end of a woman's cycle. Even romantic jealousy and reactions to emotional infidelity have been shown to fluctuate with estrogen levels (Geary *et al.*, 2002).

Vaginal absorption has been found to fluctuate with the menstrual cycle as well (Benziger & Edelson, 1983). However, few studies have been conducted and this fluctuation seems to depend on the type of substance. It is possible that because of a high volume of bacteria at the end of a menstrual cycle the absorptive capabilities of the vagina are at their lowest (Profet, 1993), and this creates an increase of depressive and anxious symptoms.

It is also possible that premenstrual syndrome constitutes an anticipatory withdrawal from semen. Since many women abstain from sexual intercourse while they are menstruating (Hedricks, 1994; Spitz, Gold, & Adams, 1975), each monthly cycle begins with an extended period of semen withdrawal for those that are having unprotected sex. The psychological symptoms commonly associated with the premenstrum could be conditioned anticipatory responses to the lack of sexual activity in the following week. This could also account for the increase of female-initiated sexual behavior pre- and postmenstrually (Bancroft et al., 1983; Stewart, 1989; Zillmann, Schweitzer, & Mundorf, 1994).

If the premenstrual syndrome were to be characterized as anticipatory semen withdrawal, then we would predict that premenstrual symptoms would occur in females who (1) had consistent unprotected sexual activity and (2) abstained from sex during their menstrual phase. Burch and Gallup (unpublished data) found that among women who never used condoms, frequency of sexual intercourse did not correlate with symptoms of premenstrual syndrome for those who had sex throughout their menstrual phase. For those who did not, intercourse frequency did correlate with premenstrual symptoms. For those who used condoms some of the time, most of the time, or all of the time (and therefore were not receiving semen on a consistent basis), frequency of sexual intercourse did not correlate with symptoms of premenstrual syndrome, regardless of whether they had sex throughout their menstrual phase. The principle variables that correlate most highly seem to be of a physical nature; for example, bloating, cramps, body aches, and nausea. This suggests that the premenstrual symptoms most affected by "semen withdrawal" may have a physiological basis.

One of the functions of the apparent antidepressant properties of semen may have evolved to fortify the bond between the female and her mate. For example, the mood-modulating effects of semen may function to enable females in commited sexual relationships to eliminate, modulate, and or regulate mood swings. Thus exposure to semen may function to enhance the commitment the female makes to her mate.

MENSTRUAL-CYCLE REGULARITY

Because so many seminal compounds have the potential to influence ovulation, we hypothesized that this combination of compounds administered frequently and consistently could affect menstrual cycle regularity. Although previous studies have examined sexual activity and its influence on cycle regularity, condom use (and therefore the presence or absence of semen in the reproductive tract) was not investigated (Burleson, Gregory, & Trevathan, 1991; Cutler *et al.*, 1979, 1985; Veith *et al.*, 1983). Burch and Gallup (unpublished data) found evidence that semen affects menstrual-cycle regularity. Intercourse frequency was correlated with cycle regularity in women who did not use condoms, but not in women who used condoms. This implies, just as with depressive symptoms, that it is not simply sexual activity *per se* that affects cycle regularity, but semen exposure.

INDUCED MENSTRUATION

Some compounds in semen have also been used to induce menstruation in human females. Although it seems contradictory that semen could induce both ovulation and menstruation, it is important to remember that males are not aware of where the female is in her cycle, and the chemistry of semen may have evolved to take advantage of any given situation. It is possible that since the compounds in semen are so small in volume, they may simply prime ongoing endocrine features of the female. If a female is nearing ovulation, the hormones in semen may be just enough to trigger the early release of the egg. If however, the female is nearing the end of her cycle, luteinizing hormone, no matter what the dose is, may have little effect. In this situation, it may be in the males' best interests to trigger menstruation, to terminate pregnancies initiated by other males (the Bruce Effect; see below). Levels of prostaglandins are relatively high at the beginning of a female's menstrual phase and the additional prostaglandins in semen may trigger menstruation. In fact, prostaglandins are commonly used as abortive agents (Creinin, 2000; Gemzell-Danielsson & Ostlund, 2000; Smith *et al.*, 2000; Wiebe, 1997). If the prostaglandin component of semen is effective in inducing menstruation in normally cycling women, we would expect this to only be found in women who do not use condoms.

Burch and Gallup (unpublished data) found that a large percentage of men (25%) reported their partner getting her period at an unexpected or inappropriate time in her cycle. A third of the males (32.2%) reported that this had occurred just after they had begun dating or during a visit. Over a quarter of females reported getting their period at an unusual or inappropriate time in their cycle. When asked if induced menstruation occurred in response to a new sex partner, or visiting an existing one, the incidence of induced menstruation rose to over a third. A greater number of females who reported induced menstruation and menstruation when visiting a partner were not using contraceptives at that time. Females who did not use contraceptives were more likely to report irregular bleeding when visiting a partner, and inappropriate menstruation when starting a new relationship.

As suggested above, it is possible that induced menstruation is, in some cases, miscarriage. If this were the case, it would constitute what is known as the Bruce Effect (Bruce, 1959), where in rodents intercourse or beginning a relationship with a new male causes the female to physiologically abort her pregnancy. An obvious implication of this is that it evolved to enable females to abort an in-pair pregnancy in favor of a higher-status extra-pair male.

COGNITION

The compounds in semen may affect more than sexual behavior or phenomena associated with the menstrual cycle. Epinephrine is released during learning and enhances memory, as is norepinephrine (Nelson, 2000). Vasopressin has been shown to enhance memory and increase mate guarding in other animals. Oxytocin has been shown to be an amnesic agent, but in some contexts can even enhance memory. Oxytocin is also implicated in bonding between individuals and increased focus and attachment to a person, such as a lover or child (Nelson, 2000). Several rodent studies have found that oxytocin is linked to monogamy (Insel & Shapiro, 1992). Opioids have been found to also trigger some amnesia, but they also increase feelings of reward in learning situations (Nelson, 2000). Estrogens have been found to enhance memory and information consolidation as do glucocorticoids in low-to-moderate doses (Nelson, 2000). Melatonin has been shown to improve short-term memory in other animals (Argyriou, Prast, & Philippu, 1998).

Given this information, it is possible that semen could have effects on female cognition, learning, and memory. In fact, Burch and Gallup (unpublished data) found, in addition to the psychological items on the Beck Depression Inventory, that women who had unprotected sex had significantly lower scores on indecisiveness and difficulty concentrating. The difficulty concentrating item showed the greatest differences across condom groups for both the Gallup *et al.* (2002) dataset and its replication (Burch & Gallup, unpublished data). Difficulty concentrating was also positively correlated with time since last intercourse among women who were having unprotected intercourse.

SPERM RETENTION: COPULATORY AND POSTCOPULATORY BEHAVIORS

Not only does sexual intercourse affect the dosage of semen, but specific behaviors during and after sex can affect retention of semen in the female reproductive tract. As a consequence of vaginal and intrauterine contractions, female orgasm during unprotected heterosexual intercourse may function to enhance sperm retention and transport (Baker & Bellis, 1995), and this could result in lower depressive symptoms.

Burch and Gallup (unpublished data) found that some postcopulatory behaviors affect depressive symptoms. Women who rarely or never used condoms showed a positive correlation between depressive symptoms and urinating after intercourse; a behavior that results in the greatest ejection or "flowback" and loss of semen according to Baker and Bellis (1995). Among these women there was also a negative correlation between depressive symptoms and lying down/cuddling with partner after intercourse, which would require the woman to remain in a supine position, aiding in the retention of semen. Thus, it would appear that postcopulatory behavior on the part of the female can affect the dose of semen she receives and females might selectively retain semen from certain males that are of higher quality (e.g. see Chapter 7 in this volume).

Working hypothesis 1: concealed ovulation and semen chemistry

It seems reasonable to suppose that in addition to ingredients that serve to maintain and support sperm, other components in semen may be a byproduct of reproductive competition during human evolutionary history. Indeed, we contend that semen chemistry has evolved in part to influence and manipulate the female reproductive system in ways that would benefit the male.

When it comes to reproduction, insemination does not always suffice to produce conception. Other things being equal, conception requires the release/exchange of gametes (sperm and eggs) under conditions that are proximate to one another in both time and space. In other words, the emphasis during evolution was on synchronizing insemination with ovulation. Among many sexually reproducing species this synchrony is achieved as a consequence of patterns of seasonal or cyclical breeding in which females often produce salient external cues that signal ovulation.

In contrast to other species, humans no longer show breeding patterns that are driven by season, cycles, or signals. Indeed, human females no longer produce ovulatory signals and have become what some people characterize as concealed ovulators (Burley, 1979). Thus the problem of synchronizing the release of semen with the release of eggs has become an important dimension of human reproductive competition.

As shown in Table 8.1, semen not only contains male hormones, but also female hormones. What are female hormones doing in human semen? Notice that follicle-stimulating hormone (FSH), luteinizing hormone (LH), and luteinizing hormone-releasing hormone are present in human semen.

In females these steroids are involved in regulating certain features of the woman's menstrual cycle. FSH causes an egg in an ovary to ripen/mature, and then an abrupt increase in LH (called the LH surge) triggers ovulation or the release of that egg. Ovulation test kits, used by women who want to get pregnant, are sensitive to the presence of LH in urine and therefore can be used to index the occurrence of ovulation. Thus it would appear that the chemistry of human semen has been selected to mimic the hormonal conditions that control ovulation, and as such may account for instances of induced ovulation (ovulation triggered by copulation at points in the menstrual cycle when ovulation would otherwise be unlikely). It is interesting to speculate that these features of semen may function as an adaptive accommodation to the fact that human females have become concealed ovulators. The loss of ovulatory cues greatly reduces the likelihood of conception as a consequence of a random sexual encounter, because there is no way to synchronize insemination with ovulation. In the absence of reliable ovulatory signals the only way to synchronize insemination with ovulation is to engage in high-frequency copulation over an extended period of time. Indeed, the reason humans practice patterns of continuous breeding (where breeding proclivity is no longer tied to season, cycles, or signals) is to accommodate reproduction under conditions in which ovulation is concealed.

In collaboration with Kenneth Gould at the Yerkes Regional Primate Research Center and Mark Wilson at Emory University, we attempted to test this hypothesis by comparing the levels of FSH and LH in human and chimpanzee semen. The rationale for undertaking this comparison was relatively simple and straightforward. Whereas chimpanzees are our closest living primate relatives, unlike women, chimpanzee females are cyclical breeders. Female chimpanzees in the ovulatory phase of their cycles show dramatic changes in genital coloration and swelling that serve to advertise or signal ovulation. As you might expect, these genital changes have a powerful effect on male chimpanzees by way of producing sexual arousal and interest. Therefore, according to our hypothesis, if the composition of human semen has evolved to make induced ovulation more likely, to compensate for the loss of ovulatory signals, then levels of FSH and LH ought to differ between human and chimpanzee semen. That is exactly what we found. Not only were levels of LH lower (and more variable) in chimpanzee than human semen, but FSH was completely missing from chimpanzee semen.

Working hypothesis 2: forced copulation and semen chemistry

From an evolutionary perspective, one of the principal problems posed to females about the prospect of rape is conception. Conception as a consequence of being raped entails very high costs. Not only does it mean that the female probably won't get any protection and/or support from the child's father, but it forfeits/precludes her opportunity to exercise mate choice. Females have a vested interest in the other 50% of the genes being carried by their offspring. Becoming pregnant if she is raped may also undermine a women's relationship with her current mate, and may impair her ability to form committed provisioning relationships with other high-quality males in the future.

Based on retrospective reports about rape, several studies have found that when victims are queried about approximately where they were in their menstrual cycles when the assault occurred, the probability of being raped appears to vary as a function of the menstrual cycle (Morgan, 1981; Rogel, 1976). Rape victims are less likely to be in the ovulatory phase than any other phase of their menstrual cycle when the assault occurs. Chavanne and Gallup (1998) reasoned that this might be a consequence of the fact that during human evolutionary history females may have been selected to unwittingly sequester themselves during mid cycle to minimize the chances of being raped at a time when conception would be most likely. To test this hypothesis, female college students were asked to respond to an anonymous survey about their menstrual cycle, and they were asked to fill out a checklist of activities they had participated in during the previous 24-hour period. These activities included a mix of low- (e.g. staying home and watching television), medium- (e.g. going to the grocery store), and high-risk behaviors (e.g. walking alone at night in a dimly lit area). Each of these activities was assigned a risk-taking score based on sets independent ratings and rankings. Then a composite risk-taking score for the previous day was derived for each respondent by dividing the sum of all of the risk-taking scores by the number of reported activities. As predicted, for females who were not taking birth-control pills there was an appreciable drop in composite risk-taking scores during the mid-phase of the menstrual cycle. These findings, showing an ovulatory-phase reduction in risk-taking behavior, have been replicated recently by Broder and Hohmann (2003).

As an extension of these results, Petralia and Gallup (2002) tested female college students for differences in their ability to resist and/or deter a forced sexual encounter as a function of where they happened to be in their menstrual cycles. In addition to being asked anonymous questions about their menstrual cycles, subjects were also tested using ovulation test kits that are sensitive to urinary levels of LH to pinpoint those who were in the ovulatory phase. Grip strength, using a hand-held dynamometer, was taken as an indirect measure of resistance. Following a grip-strength pretest, subjects read either a sexual assault scenario or a sexually neutral script and were then tested again. The only respondents that showed a significant increase in grip strength from the pretest to the test trial were those who read the sexual assault passage and were in the ovulatory phase of their menstrual cycle. Women in all the other conditions showed a decrease in grip strength from the pretest to the test. Thus in addition to changes in risk-taking behavior that occur during the mid-phase of the menstrual cycle, there may be corresponding contextual/situational changes in the ability to resist being raped as a function of the risk of conceiving.

Does the likelihood of conception vary as a function of whether sex is forced or consensual? There is evidence that the risk of becoming pregnant as a result of being raped is approximately three times higher than as a consequence of having a consensual sexual encounter (see review by Gottschall & Gottschall, 2003). In light of the fact that females appear to behave in ways to minimize conception as a byproduct of being raped, the increased risk of conception is even more anomalous. Therefore, our second working hypothesis is that there may be mechanisms operating at the level of the testicles that adjust certain features of semen chemistry to make induced ovulation more likely as a consequence of sexual assault.

There are several ways to test this hypothesis. Since semen samples are routinely collected from rape victims, one approach would be to assay these samples to see if the levels of FSH and LH are higher than in those obtained from women who have engaged in consensual encounters. Another test of this hypothesis would be to collect semen samples from donors who are exposed to pornography containing different levels of themes related to rape and/or violence. According to our hypothesis, levels of donor FSH and LH would be expected to increase as rape/violence themes become more prominent in pornography that sperm donors watch while they masturbate.

Working hypothesis 3: semen chemistry and sexual orientation

In addition to the vaginal application of semen, there is reason to believe that the oral ingestion or anal application of semen may also produce comparable effects. Indeed, a recent study by Koelman et al. (2000) shows that the incidence of preeclampsia among pregnant women varies as a function of oral sex. Not only do pregnant women who have oral sex with their partner have lower rates of preeclampsia, those that ingest their partner's semen derive an even greater protective effect. As shown by the effects of oral contraceptives, many steroids survive the ingestion process and therefore could continue to have important endocrine/reproductive consequences (testosterone may be an exception).

As shown in Table 8.1, the presence of oxytocin in semen has some interesting implications in its own right. In females, oxytocin release is associated with three conditions: childbirth, breastfeeding, and orgasm (Nelson, 2000). Given the growing evidence that oxytocin serves to enhance social and attachment bonds, oxytocin release at childbirth and during breastfeeding obviously functions to facilitate the development of the maternal–infant bond which is crucial for the infant's survival and well being. Likewise, oxytocin release at orgasm would serve to enhance and maintain the female's attachment and affectional bond to her male partner. The presence of oxytocin in semen would appear to converge with oxytocin release at orgasm to promote the same effect.

This has interesting implications for the development of sexual orientation. Whereas most males, unlike females, receive little or no exposure to oxytocin, semen exchange often occurs through oral and anal sex among homosexual males. Indeed, we would predict that because of the effect of oxytocin on the development of attachment and affectional bonds, the likelihood of developing a homosexual orientation as a consequence of an early same-sex sexual encounter among males ought to vary as a function of semen exchange. In other words, if the initial encounter involves the use of condoms then the likelihood of a same-sex sexual orientation would be less.

In spite of the risks associated with spreading and/or contracting the AIDS virus, the evidence suggests that a substantial proportion of homosexual males continue to practice unsafe sex; that is, abstain from using condoms (Halkitis, Parsons, & Wilton, 2003). Reminiscent of the apparent antidepressant effects of semen exposure on females (Gallup et al.,

2002), reports of enhanced emotional and psychological effects of unprotected anal intercourse among homosexual males are widespread (Bancroft *et al.*, 2003; Halkitis & Parsons, 2003; Scarce, 1999). Indeed, there is evidence that following anal insemination homosexual males often attempt to retain the semen (by lying on their backs or through the use of "butt plugs") for extended periods of time (Scarce, 1999).

References

Abbaticchio, G., Giorgino, R., Urago, M., *et al.* (1981). Hormones in the seminal plasma. Cortisol. *Acta Europaea Fertilitatis*, **12**(3), 239–44.

Abdullah, Y. H. and Hamadah, K. (1975). Effect of ADP on PGE1 formation in the blood platelets from patients with depression, mania and schizophrenia. *British Journal of Psychiatry*, **127**, 591–5.

Aiman, J., McAsey, M., and Harms, L. (1988). Serum and seminal plasma prolactin concentrations in men with normospermia, oligospermia, or azoospermia. *Fertility and Sterility*, **49**(1), 133–7.

Allen, S. S., Hatsukami, D., Christianson, D., and Nelson, D. (1996). Symptomatology and energy intake during the menstrual cycle in smoking women. *Journal of Substance Abuse*, **8**, 303–19.

Argyriou, A., Prast, H., and Philippu, A. (1998). Melatonin facilitates short-term memory. *European Journal of Pharmacology*, **349**(2–3), 159–62.

Asch, R. H., Fernandez, E. O., Siler-Khodr, T. M., and Pauerstein, C. J. (1984). Peptide and steroid hormone concentrations in human seminal plasma. *International Journal of Fertility*, **29**(1), 25–32.

Bachmann, G. A., Leiblum, S. R., Kemmann, E., *et al.* (1984). Sexual expression and its determinants in the post-menopausal woman. *Maturitas*, **6**, 19–29.

Baker, R. R. and Bellis, M. A. (1995). *Human Sperm Competition: Copulation, Masturbation, and Infidelity*. London: Chapman and Hall.

Bancroft, J., Sanders, D., Davidson, D., and Warner, P. (1983). Mood, sexuality, hormones, and the menstrual cycle, III. Sexuality and the role of androgens. *Psychosomatic Medicine*, **45**, 509–16.

Bancroft, J., Janssen, E., Strong, D., *et al.* (2003). Sexual risk taking in gay men: the relevance of sexual arousability, mood, and sensation seeking. *Archives of Sexual Behavior*, **32**, 555–73.

Benziger, D. P. and Edelson, J. (1983). Absorption from the vagina. *Drug Metabolism Reviews*, **14**, 137–68.

Bloch, M., Schmidt, P. J., and Rubinow, D. R. (1997). Premenstrual syndrome: evidence for symptom stability across cycles. *American Journal of Psychiatry*, **154**, 1741–6.

Broder, A. and Hohmann, N. (2003). Variations in risk taking behavior over the menstrual cycle: an improved replication. *Evolution and Human Behavior*, **24**(6), 391–8.

Brotherton, J. (1990a). Cortisol and transcortin in human seminal plasma and amniotic fluid as estimated by modern specific assays. *Andrologia*, **22**(3), 197–204.

Brotherton, J. (1990b). Ferritin: another pregnancy-specific protein in human seminal plasma and amniotic fluid, as estimated by six methods. *Andrologia*, **22**(6), 597–607.

Brotherton, J. (1990c). Vasopressin: another pregnancy protein in human seminal plasma. *Andrologia*, **22**(4), 305–7.

Brown, R. P., Kocsis, J. H., Caroff, S., *et al.* (1987). Depressed mood and reality disturbance correlate with decreased nocturnal melatonin in depressed patients. *Acta Psychiatrica Scandinavica*, **76**(3), 272–5.

Bruce, H. (1959). An exteroceptive block to pregnancy in the mouse. *Nature*, **182**, 105.

Burleson, M. H., Gregory, W. L., and Trevathan, W. R. (1991). Heterosexual activity and cycle length variability: effect of gynecological maturity. *Physiology and Behavior*, **50**, 863–6.

Burley, N. (1979). The evolution of concealed ovulation. *American Naturalist*, **114**, 835–8.

Burt, V. K., Altschuler, L. L., and Rasgon, N. (1998). Depressive symptoms in the perimenopause: prevalence, assessment, and guidelines for treatment. *Harvard Review of Psychiatry*, **6**, 121–32.

Cameron, O. G., Lee, M. A., Kotun, J., and Murphy, S. T. (1986). Circadian fluctuations in anxiety disorders. *Biological Psychiatry*, **21**, 567–8.

Cameron, O. G., Kuttesch, D., McPhee, K., and Curtis, G. C. (1988). Menstrual fluctuation in the symptoms of panic anxiety. *Journal of Affective Disorders*, **15**, 169–74.

Carter, C. S. (1992). Hormonal influences on human sexual behavior. In J. B. Becker, S. M. Breedlove, and D. Crews, eds., *Behavioral Endocrinology*. Cambridge, MA: MIT Press, pp. 131–42.

Carter, C. S. and Altemus, M. (1997). Integrative functions of lactational hormones in social behavior and stress management. *Annual New York Academy of Science*, **807**, 164–74.

Cassone, V. M., Warren, W. S., Brooks, D. S., and Lu, J. (1993) Melatonin, the pineal gland, and circadian rhythms. *Journal of Biological Rhythms*, **8**, Suppl, S73–81.

Chan, S. Y. and Tang, L. C. (1983). Immunoreactive LHRH-like factor in human seminal plasma. *Archives of Andrology*, **10**(1), 29–32.

Chavanne, T. J., and Gallup, G. G., Jr. (1998). Variation in risk taking behavior among female college students as a function of the menstrual cycle. *Evolution and Human Behavior*, **19**, 27–31.

Cook, B. L., Noyes, R., Garvey, M. J., and Beach, V. (1990). Anxiety and the menstrual cycle in panic disorder. *Journal of Affective Disorders*, **19**, 221–6.

Coope, J. (1996). Hormonal and non-hormonal interventions for menopausal symptoms. *Maturitas*, **23**, 159–68.

Coope, J., Thomson, J. M., and Poller, L. (1975). Effects of "natural oestrogen" replacement therapy on menopausal symptoms and blood clotting. *British Medical Journal*, **4**(5989), 139–43.

Creinin, M. D. (2000). Medical abortion regimens, historical context and overview. *American Journal of Obstetrics and Gynecology*, **183**(2 Suppl), S3–9.

Cutler, W., Garcia, C., and Krieger, A. (1979). Luteal phase defects: a possible relationship between short hyperthermic phase and sporadic sexual behavior in women. *Hormones and Behavior*, **13**, 214–18.

Cutler, W., Preti, G., Huggins, G., Erickson, B., and Garcia, C. (1985). Sexual behavior frequency and biphasic ovulatory type menstrual cycles. *Physiology and Behavior*, **34**, 805–10.

Davidson, A., Vermesh, M., Paulson, R. J., Graczykowski, J. W., and Lobo, R. A. (1989). Presence of immunoreactive beta-endorphin and calcitonin in human seminal plasma, and their relation to sperm physiology. *Fertility and Sterility*, **51**(5), 878–80.

Davis, F. C. (1997). Melatonin: role in development. *Journal of Biological Rhythms*, **12**(6), 498–508.

de Medeiros, S. F., Amato, F., Bacich, D., *et al.* (1992). Distribution of the beta-core human chorionic gonadotrophin fragment in human body fluids. *Journal of Endocrinology*, **135**(1), 175–88.

Dennerstein, L., Gotts, G., Brown, J. B., and Morse, C. A. (1994). The relationship between the menstrual cycle and female sexual interest in women with PMS complaints and volunteers. *Psychoneuroendocrinology*, **19**, 293–304.

Derzko, C. M (1990). Role of danazol in relieving the premenstrual syndrome. *Journal of Reproductive Medicine*, **35**(1) Suppl, 97–102.

Dixson, A. F. (1987). Effects of andrenalectomy upon proceptivity, receptivity and sexual attractiveness in ovariectomized marmosets (*Callithrix jacchus*). *Physiology and Behavior*, **39**(4), 495–9.

Doss, S. H. and Louca, N. A. (1991). Semen finger print. *Forensic Science International*, **51**(1), 1–12.

Fait, G., Vered, Y., Yogev, L., *et al.* (2001). High levels of catecholamines in human semen: a preliminary study. *Andrologia*, **33**(6), 347–50.

Fernandez, D., Valdivia, A., Irazusta, J., Ochoa, C., and Casis L. (2002) Peptidase activities in human semen. *Peptides*, **23**(3), 461–8.

Gallup, G. G., Jr., Burch, R. L., and Platek, S. (2002). Does semen contain antidepressant properties? *Archives of Sexual Behavior*, **39**(3), 289–91.

Geary, D. C., DeSoto, M. C., Hoard, M. K., Sheldon, M., and Cooper, L. (2002). Estrogens and relationship jealousy. *Human Nature*, **12**, 299–320.

Gemzell-Danielsson, K. and Ostlund, E. (2000). Termination of second trimester pregnancy with mifepristone and gemeprost: the clinical experience of 197 consecutive cases. *Acta Obstetrica Gynecologie Scandinavica*, **79**, 702–6.

Gkonos, P. J., Kwok, C. K., Block, N. L., and Roos, B. A. (1994). Identification of the human seminal TRH-like peptide pGlu-Phe-Pro-NH2 in normal human prostate. *Peptides*, **15**(7), 1281–3.

Golden, R. N., Heine, A. D., Ekstrom, R. D., *et al.* (2002). A longitudinal study of serotonergic function in depression. *Neuropsychopharmacology*, **26**(5), 653–9.

Golub, S. (1976). The magnitude of premenstrual anxiety and depression. *Psychosomatic Medicine*, **38**, 4–12.

Gonzales, G. F. and Garcia-Hjarles, M. A. (1990). Blood/seminal serotonin levels in infertile men with varicocele. *Archives of Andrology*, **24**(2), 193–9.

Gonzales, G. F., Garcia-Hjarles, M. A., Napuri, R., Coyotupa, J., and Guerra-Garcia, L. (1989). Blood serotonin levels and male infertility. *Archives of Andrology*, **22**(1), 85–9.

Gottschall, J. A. and Gottschall, T. A. (2003). Are per-incident rape-pregnancy rates higher than per-incident consensual pregnancy rates? *Human Nature*, **14**(1), 1–20.

Goverde, H. J., Bisseling, J. G., Wetzels, A. M., *et al.* (1998). A neuropeptide in human semen: Oxytocin. *Archives of Andrology*, **41**(1), 17–22.

Grattan, D. R. (2001). The actions of prolactin in the brain during pregnancy and lactation. *Progress in Brain Research*, **133**, 153–71.

Guay, A. T. (2001). Decreased testosterone in regularly menstruating women with decreased libido: a clinical observation. *Journal of Sex and Marital Therapy*, **27**(5), 513–19.

Halkitis, P. N. and Parsons, J. T. (2003). Intentional unsafe sex (barebacking) among HIV positive gay men who seek sexual partners on the Internet. *AIDS Care*, **15**(3), 367–78.

Halkitis, P. N., Parsons, J. T., and Wilton, L. (2003). Barebacking among gay and bisexual men in New York City: explanations for the emergence of intentional unsafe behavior. *Archives of Sexual Behavior*, **32**(4), 351–8.

Hampl, R., Hill, M., Sterzl, I., and Starka, L. (2000). Immunomodulatory 7-hydroxylated metabolites of dehydroepiandrosterone are present in human semen. *Journal of Steroid Biochemistry and Molecular Biology*, **75**(4–5), 273–6.

Hedricks, C. A. (1994). Female sexual activity across the human menstrual cycle. *Annual Review of Sex Research*, **5**, 122–72.

Henderson, B. J. and Whissell, C. (1997). Changes in women's emotions as a function of emotion valence, self determined category of premenstrual distress, and day in the menstrual cycle. *Psychological Reports*, **80**, 1272–4.

Hendrick, V., Altshuler, L. L., and Suri, R. (1998). Hormonal changes in the postpartum and implications for postpartum depression. *Psychosomatics*, **39**(2), 93–101.

Hepburn, D. A., Deary, I. J., MacLeod, K. M., and Frier, B. M. (1996). Adrenaline and psychometric mood factors: a controlled case study of two patients with bilateral adrenalectomy. *Personality and Individual Differences*, **20**(4), 451–5.

Homberg, M. and Samuelsson, B. (1966). Prostaglandins in human seminal plasma: prostaglandins and related factors. *Journal of Biological Chemistry*, **241**, 257–63.

Insel, T. R. and Shapiro, L. E. (1992). Oxytocin receptors and maternal behavior. In C. A. Pedersen and J. D. Caldwell, eds., *Oxytocin in Maternal, Sexual, and Social Behaviors*. New York: New York Academy of Sciences, pp. 122–41.

Jarett, L. (1984). Psychosocial and biological influences on menstruation: synchrony, cycle length, and regularity. *Psychoneuroendocrinology*, **9**, 21–8.

Jarvis, T. J. and McCabe, M. P. (1991). Women's experience of the menstrual cycle. *Journal of Psychosomatic Research*, **35**, 651–60.

Josefsson, A., Berg, G., Nordin, C., and Sydsjo, G. (2001). Prevalence of depressive symptoms in late pregnancy and postpartum. *Acta Obstetricia et Gynecologica Scandinavica*, **80**, 251–5.

Keller, P. J., Riedmann, R., Fischer, M., and Gerber, C. (1981). Oestrogens, gonadotropins and prolactin after intra-vaginal administration of oestriol in post-menopausal women. *Maturitas*, **3**(1), 47–53.

Kelly, R. W. (1995). Immunosuppressive mechanisms in semen: implications for contraception. *Human Reproduction*, **10**(7), 1686–93.

Khan, Z. and Smyth, D. G. (1993). Isolation and identification of N-terminally extended forms of 5-oxoprolylglutamylprolinamide (Glp-Glu-Pro-NH2), a thyrotropin-releasing-hormone (TRH)-like peptide present in human semen. *European Journal of Biochemistry*, **212**(1), 35–40.

Khan, Z., Aitken, A., Garcia, J. R., and Smyth, D. G. (1992). Isolation and identification of two neutral thyrotropin releasing hormone-like peptides, pyroglutamyl-phenylalanineproline amide and pyroglutamylglutamineproline amide, from human seminal fluid. *Journal of Biological Chemistry*, **267**(11), 7464–9.

Klaiber, E. L., Broverman, D. M., Vogel, W., and Kobayashi, Y. (1979). Estrogen therapy for severe persistent depressions in women. *Archives of General Psychiatry*, **36**(5), 550–4.

Koelman, C. A., Coumans, A. B., Nijam, H. W., *et al.* (2000). Correlation between oral sex and a low incidence of preeclampsia: a role for soluble HLA in seminal fluid? *Journal of Reproductive Immunology*, **46**(2), 155–66.

Kwan, T. K., Trafford, D. J., Makin, H. L., Mallet, A. I., and Gower, D. B. (1992). GC-MS studies of 16-androstenes and other C19 steroids in human semen. *Journal of Steroid Biochemistry and Molecular Biology*, **43**(6), 549–56.

Lawrie, T. A., Herxheimer, A., and Dalton, K. (2000). Oestrogens and progestens for preventing and treating postnatal depression. *Cochrane Database of Systematic Reviews*, (2), CD001690.

Lebowitz, B. D., Pearson, J. L., Schneider, L. S., *et al.* (1997). Diagnosis and treatment of depression in late life: consensus statement update. *Journal of the American Medical Association*, **278**(14), 1186–90.

Lebowitz, B. D., Pollock, B. G., and Schneider, L. S. (1997). Estrogen in geriatric psychopharmacology. *Psychopharmacology Bulletin*, **33**(2), 287–8.

Lee, J. N., Lian, J. D., Lee, J. H., and Chard, T. (1983). Placental proteins (human chorionic gonadotropin, human placental lactogen, pregnancy-specific beta 1-glycoprotein, and placental protein 5) in seminal plasma of normal men and patients with infertility. *Fertility and Sterility*, **39**(5), 704–6.

Lewy, A. J., Wehr, T. A., Goodwin, F. K., Newsome, D. A., and Rosenthal, N. E. (1984). Manic-depressive patients may be supersensitive to light. *Lancet*, **1**(8216), 383–4.

Loumaye, E., De Cooman, S., and Thomas, K. (1980). Immunoreactive relaxin-like substance in human seminal plasma. *Journal of Clinical Endocrinology and Metabolism*, **50**(6), 1142–3.

Luboshitzky, R., Shen-Orr, Z., and Herer, P. (2002). Seminal plasma melatonin and gonadal steroids concentrations in normal men. *Archives of Andrology*, **48**(3), 225–32.

Mackenzie, I. Z., Bradley, S., and Mitchell, M. D. (1980). Prostaglandin levels on cord venous plasma at delivery related to labor. *Advances in Prostaglandin and Trombocic Research*, **8**, 1401–5.

MacLennan, A. H. (1991). The role of the hormone relaxin in human reproduction and pelvic girdle relaxation. *Scandinavian Journal of Rheumatology Supplement*, **88**, 7–15.

Maegawa, M., Kamada, M., Irahara, M., *et al.* (2002). A repertoire of cytokines in human seminal plasma. *Journal of Reproductive Immunology*, **54**(1–2), 33–42.

Matteo, S. (1987). The effect of job stress and job interdependency on menstrual cycle length, regularity, and synchrony. *Psychoneuroendocrinology*, **12**, 467–76.

Matteo, S. and Rissman, E. F. (1984). Increased sexual activity during the midcycle portion of the human menstrual cycle. *Hormones and Behavior*, **18**, 249–55.

McCoy, N., Cutler, W., and Davidson J. M. (1985). Relationships among sexual behavior, hot flashes, and hormone levels in perimenopausal women. *Archives of Sexual Behavior*, **14**, 385–94.

Menchini-Fabris, G. F., Canale, D., Izzo, P. L., Olivieri, L., and Bartelloni, M. (1984). Free L-carnitine in human semen: its variability in different andrologic pathologies. *Fertility and Sterility*, **42**(2), 263–7.

Meyers, B. S. and Moline, M. L. (1997). The role of estrogen in late life depression: opportunities and barriers to research. *Psychopharmacology Bulletin*, **33**, 289–91.

Mondina, R., Chiara, F., Aspesi, F., and Polvani, F. (1976). Pituitary hormones in human seminal plasma. *Sperm Action Progress in Reproductive Biology*, **1**, 121–4.

Morgan, J. B. (1981). *Relationship Between Rape and Physical Damage during Rape and Phase of Sexual Cycle during which Rape Occurred*. Doctoral dissertation, University of Texas at Austin.

Morris, N. M., Udry, J. R., Khan Dawood, F., and Dawood, M. Y. (1987). Marital sex frequency and midcycle female testosterone. *Archives of Sexual Behavior*, **16**(1), 27–37.

Mungan, N. A., Mungan, G., Basar, M. M., Baykam, M., and Atan, A. (2001). Effect of seminal plasma calcitonin levels on sperm mobility. *Archives of Andrology*, **47**(2), 113–17.

Nelson, R. (2000). *An Introduction to Behavioral Endocrinology*. Sunderland, MA: Sinauer Associates.

Ney, P. G. (1986). The intravaginal absorption of male generated hormones and their possible effect on female behavior. *Medical Hypotheses*, **20**, 221–31.

Nygren, K. G. and Rybo, G. (1983). Prostaglandins and menorrhagia. *Acta Obstetrica Gynecologie Scandinavica Supplemental*, **113**, 101–3.

Pekary, A. E., Hershman, J. M., and Friedman, S. (1983). Human semen contains thyrotropin releasing hormone (TRH), a TRH-homologous peptide, and TRH-binding substances. *International Journal of Andrology*, **4**(6), 399–407.

Pekary, A. E., Reeve, J. R., Jr., Smith, V. P., and Friedman, S. (1990). In-vitro production of precursor peptides for thyrotropin-releasing hormone by human semen. *International Journal of Andrology*, **13**(3), 169–79.

Petralia, S. M., and Gallup, G. G., Jr. (2002). Effects of a sexual assault scenario on handgrip strength across the menstrual cycle. *Evolution and Human Behavior*, **23**(1), 3–10.

Preti, G., Cutler, W. B., Garcia, C. R., and Huggins, G. (1986). Human axillary secretions influence women's menstrual cycles: the role of donor extract of females. *Hormones and Behavior*, **20**(4): 474–82.

Profet, M. (1993). Menstruation as a defense against pathogens transported by sperm. *Quarterly Review of Biology* **68**, 335–86.

Purvis, K., Landgren, B. M., Cekan, Z., and Diczfalusy, E. (1975). Indices of gonadal function in the human male. II. Seminal plasma levels of steroids in normal and pathological conditions. *Clinical Endocrinology*, **4**(3): 247–58.

Rigg, L. A., Milanes, B., Villanueva, B., and Yen, S. C. C. (1977). Efficacy of intravaginal and intranasal administration of micronized estradiol 17-beta. *Journal of Clinical Endocrinology and Metabolism*, **45**: 1261–4.

Rogel, M. J. (1976). *Biosocial Aspects of Rape*. Doctoral dissertation, University of Chicago.

Roy-Byrne, P., Rubinow, D., Gold, P., and Post, R. (1984). Possible antidepressant effects of oral contraceptives: case report. *Journal of Clinical Psychiatry*, **45**, 350–2.

Ruiz-Pesini, E., Alvarez, E., Enríquez, J. A., and López-Pérez, M. J. (2001). Association between seminal plasma carnitine and sperm mitochondrial enzymatic activities. *International Journal of Andrology*, **24**(6), 335–40.

Sandberg, F., Ingelman-Sundberg, A., Ryden, G., and Joelsson, I. (1968). The absorption of tritium labeled prostaglandin E1 from the vagina of non-pregnant women. *Acta Obstetrica et Gynecologica Scandinavica*, **47**, 22–6.

Sastry, B. V., Janson, V. E., and Owens, L. K. (1991). Significance of substance P- and enkephalin-peptide systems in the male genital tract. *Annals of the New York Academy of Sciences*, **632**, 339–53.

Scarce, M. (1999). A ride on the wild side. *POZ*, (2), 52–5, 70–1.

Schiff, I., Tulchinsky, D., and Ryan, K. J. (1977). Vaginal absorption of estrone and 17-beta estradiol. *Fertility and Sterility*, **28**, 1063–6.

Schneider, L. S., Small, G. W., Hamilton, S. H., *et al.* (1997). Estrogen replacement and response to fluoxetine in a multicenter geriatric depression trial. *American Journal of Geriatric Psychiatry*, **5**(2), 97–106.

Seppälä, M., Koskimies, A. I., Tenhunen, A., *et al.* (1985). Pregnancy proteins in seminal plasma, seminal vesicles, preovulatory follicular fluid, and ovary. *Annals of the New York Academy of Sciences*, **442**, 212–26.

Sherwin, B. B., and Gelfand, M. M. (1985). Sex steroids and affect in the surgical menopause: a double blind, cross over study. *Psychoneuroendocrinology* **10**(3), 325–35.

Sheth, A. R., Shah, G. V., and Mugatwala, P. P. (1976). Levels of luteinizing hormone in semen of fertile and infertile men and possible significance of luteinizing hormone in sperm metabolism. *Fertility and Sterility*, **27**(8), 933–6.

Singer, R., Bruchis, S., Sagiv, M., *et al.* (1989). Beta-endorphin and calcitonin in human semen. *Archives of Andrology*, **23**(1): 77–81.

Skibinski, G., Kelly, R. W., Harrison, C. M., McMillan, L. A., and James, K. (1992). Relative immunosuppressive activity of human seminal prostaglandins. *Journal of Reproductive Immunology*, **22**(2), 185–95.

Smith, B. B., Timm, K. I., Reed, P. J., and Christensen, M. (2000). Use of cloprostenol as an abortifacient in the llama (*Lama glama*). *Theriogenology*, **54**, 497–505.

Souêtre, E., Salvati, E., Belugou, J. L., *et al.* (1987). Seasonality of suicides: environmental, sociological and biological covariations. *Journal of Affective Disorders*, **13**(3), 215–25.

Spitz, C. J., Gold, A. R., and Adams, D. B. (1975). Cognitive and hormonal factors affecting coital frequency. *Archives of Sexual Behavior*, **4**, 249–63.

Stahl, S. M. (2001). *Essential Psychopharmacology: Neuroscientific Basis and Practical Applications*. Cambridge: Cambridge University Press.

Stern, K. and McClintock, M. (1998). Regulation of ovulaton by human pheromones. *Nature*, **392**, 177–9.

Stewart, D. E. (1989). Positive changes in the premenstrual period. *Acta Psychiatrica Scandinavica*, **79**, 400–5.

Stewart, D. R., Celniker, A. C., Taylor, C. A., Jr., *et al.* (1990). Relaxin in the peri-implantation period. *Journal of Clinical Endocrinology and Metabolism*, **70**(6), 1771–3.

Trevathan, W., Burleson, M., and Gregory, W. (1993). No evidence for menstrual synchrony in lesbian couples. *Psychoneuroendocrinology*, **18**, 425–35.

Turner, R. A., Altemus, M., Enos, T., Cooper, B., and McGuinness, T. (1999). Preliminary research on plasma oxytocin in normal cycling women: investigating emotion and interpersonal distress. *Psychiatry*, **62**(2), 97–113.

Utian, W. H. (1972). The true clinical features of postmenopause and oophorectomy, and their response to oestrogen therapy. *South African Medical Journal*, **46**, 732–7.

Valsa, J., Skandhan, K. P., and Umarvanshi, V. (1992). Cholesterol in normal and pathological seminal plasma. *Panminerva Medica*, **34**(4), 160–2.

Van Goozen, S. H. M., Wiegant, V. M., Endert, E., and Helmond, F. A. (1997). Psychoendocrinological assessment of the menstrual cycle: the relationship between hormones, sexuality, and mood. *Archives of Sexual Behavior*, **26**(4), 359–82.

van Overveld, F. W., Haenen, G. R., Rhemrev, J., Vermeiden, J. P., and Bast, A. (2000). Tyrosine as important contributor to the antioxidant capacity of seminal plasma. *Chemical and Biological Interactions*, **127**(2), 151–61.

Veith, J., Buck, M., Getzlaf, S., Van Dalfsen, P., and Slade, S. (1983). Exposure to men influences the occurrence of ovulation in women. *Physiology and Behavior*, **31**, 313–15.

Villanueva, B., Casper, R., and Yen, S. S. C. (1981). Intravaginal administration of progesterone: enhanced absorption after estrogen treatment. *Fertility and Sterility*, **35**, 433–7.

von Sydow, K. (1999). Sexuality during pregnancy and after childbirth, a metacontent analysis of 59 studies. *Journal of Psychosomatic Research*, **47**, 27–49.

Weaver, D. R. (1997). Reproductive safety of melatonin: a "wonder drug" to wonder about. *Journal of Biological Rhythms*, **12**(6), 682–9.

Weiss, G. (1995) Relaxin used to produce the cervical ripening of labor. *Clinical Obstetrics and Gynecology*, **38**(2), 293–300.

Weller, A. and Weller, L. (1998). Prolonged and very intensive contact may not be conducive to menstrual synchrony. *Psychoneuroendocrinology*, **23**, 19–32.

Wester, R. C., Noonan, P. K., and Maibach, H. I. (1980). Variations in percutaneous absorption of testosterone in the rhesus monkey due to anatomic site of application and frequency of application. *Archives of Dermatological Research*, **267**, 229–35.

Wiebe, E. R. (1997). Choosing between surgical abortions and medical abortions induced with methotrexate and misoprostol. *Contraception*, **55**, 67–71.

Williams, K. E. and Koran, L. M. (1997). Obsessive-compulsive disorder in pregnancy, the peurperium, and the premenstruum. *Journal of Clinical Psychiatry*, **58**, 330–4.

Williams, T. J. and Krahenbuhl, G. S. (1997). Menstrual cycle phase and running economy. *Medicine and Science in Sports and Exercise*, **29**, 1609–18.

Winslow, J., Hastings, N., Carter, C., Harbaugh, C., and Insel, T. (1993). A role for central vasopressin in pair bonding in monogamous prairie voles. *Nature*, **365**, 545–8.

Yamazaki, I. (1984). Serum concentration patterns of an LHRH agonist, gonadotrophins and sex steroids after subcutaneous, vaginal, rectal and nasal administration of the agonist to pregnant rats. *Journal of Reproduction and Fertility*, **72**(1), 129–36.

Yie, S. M., Daya, S., Brown, G. M., Deys, L., and Young Lai, E. V. (1991). Melatonin and aromatase stimulating activity of human seminal plasma. *Andrologia*, **23**(3), 227–31.

Zalata, A., Hafez, T., Van Hoecke, M. J., and Comhaire, F. (1995). Evaluation of beta-endorphin and interleukin-6 in seminal plasma of patients with certain andrological diseases. *Human Reproduction*, **10**(12), 3161–5.

Zillmann, D., Schweitzer, K. J., and Mundorf, N. (1994). Menstrual cycle variation of women's interest in erotica. *Archives of Sexual Behavior*, **23**, 579–97.

Zweifel, J. E. and O'Brien, W. H. (1997). A meta analysis of the effect of hormone replacement therapy upon depressed mood. *Psychoneuroendocrinology*, **22**(3), 189–212.

9

Mate retention, semen displacement, and sperm competition in humans

AARON T. GOETZ AND TODD K. SHACKELFORD
Florida Atlantic University

Introduction

Competition between males to fertilize a female's eggs can occur before, during, and after copulation (Parker, 1970; and see Birkhead & Møller, 1998). When the sperm of two or more males simultaneously occupies the reproductive tract of a female and competes to fertilize her eggs, sperm competition occurs (Parker, 1970). Sperm competition has been documented or inferred to exist in many species, ranging from molluscs (Baur, 1998) and insects (Simmons, 2001) to birds (Birkhead & Møller, 1992) and humans (Baker & Bellis, 1993a, 1993b; Gallup et al., 2003; Shackelford, 2003; Shackelford et al., 2002, 2004; Smith, 1984; Wyckoff, Wang, & Wu, 2000).

For species that practice social monogamy, the mating system in which males and females form long-term pair bonds but also pursue extra-pair copulations (e.g. most birds and humans), female sexual infidelity creates the primary context for sperm competition (Birkhead & Møller, 1992; Smith, 1984). Males of such species may have adaptations that decrease the likelihood that a rival male's sperm will fertilize his partner's eggs – adaptations that decrease the likelihood of being cuckolded, unwittingly investing resources in genetically unrelated offspring. Male sexual jealousy, for example, is one of the most widely researched human anti-cuckoldry adaptations. Male sexual jealousy is hypothesized to motivate men to deter a mate from a sexual infidelity or a permanent defection from the mateship, and to deter rivals from mate poaching (e.g. Buss et al., 1992; Daly, Wilson, & Weghorst, 1982; Harris, 2003; Symons, 1979; White & Mullen, 1989). Others have described more specific adaptations that may combat sperm competition. Baker and Bellis (1993a), for example, demonstrated that men may have physiological adaptations that function to increase the likelihood that their sperm will out-compete rival sperm to fertilize their partner's eggs. In a study of couples in committed, sexual relationships, Baker and Bellis (1993a) documented that, at the couple's next copulation, men insemi-nated more sperm when the couple had spent a lesser proportion of their time together since their last copulation. As the proportion of time together decreases, the likelihood of female infidelity increases, creating a higher risk of sperm competition (Baker & Bellis, 1993a, 1995). Inseminating more sperm following a separation may function to outnumber or "flush out" rival sperm that may be present in the reproductive tract of the female (Baker & Bellis, 1993a; Parker, 1970).

This temporally variable risk of sperm competition produces specific physio-logical responses apparently designed to "correct" any female sexual infidelity that might have occurred while the couple was separated. Some men, however, may be mated to women who *recurrently* place them at a high risk of sperm competition. Female physical attractiveness and certain personality character-istics that attract rival men, for example, may increase the likelihood of female sexual infidelity and therefore place her partner at a high recurrent risk of sperm competition. Because a woman's physical attractiveness indexes her reproductive value and fertility (Singh, 1993; Symons, 1979), physically attrac-tive women are desired partners for long-term, short-term, and extra-pair mate-ships (Buss, 1989; Buss & Schmitt, 1993; Li *et al.*, 2002; Regan *et al.*, 2000). Accordingly, physically attractive women are more likely to have had men try to poach them away from their current partners (Schmitt & Buss, 2001), and men married to more physically attractive women devote more effort to retain-ing their mates (Buss & Shackelford, 1997).

Direct evidence that physically attractive women are more likely to commit infidelity comes from research examining women's waist-to-hip ratios (WHRs) and their sexual behavior. WHR is a key component of female physical attrac-tiveness (Dijkstra & Buunk, 2001; Singh, 1993; Streeter & McBurney, 2003). Low WHR is judged cross-culturally to be physically attractive, perhaps because it is a reliable indicator of reproductive age, sex hormone profile, and disease resistance – features associated with health and fertility (Singh, 1993). Hughes and Gallup (2003) documented that women with low WHR reported commit-ting more infidelities and having more sexual partners than women with high WHR. Thus, because physically attractive women attract more mate poachers and commit more infidelities, they may put their partners at a high recurrent risk of sperm competition.

Another set of factors that may place a man at a high recurrent risk of sperm competition is his partner's personality traits. The five-factor model of person-ality describes five dimensions of stable individual differences in personality

(Surgency, Agreeableness, Conscientiousness, Emotional Stability, and Openness to Experience; Goldberg, 1982; Norman, 1963). Schmitt and Buss (2001) found positive and significant relationships between a woman's Surgency and Openness to Experience and the likelihood of having had someone try to poach her away from an existing mateship. Similarly, Gangestad and Simpson (1990) found that women who are more socially dominant and extraverted (high in Surgency) are significantly more willing to have sex without indicators of commitment and emotional closeness. Sexual promiscuity, in turn, is a good predictor of infidelity (Hughes & Gallup, 2003). Women who are higher in Surgency and Openness to Experience, therefore, are more often given the opportunity to engage in extra-pair copulations. No data exist on the relationship between Surgency and Openness to Experience and the actual occurrence of infidelity, but because infidelity can only occur if the opportunity for infidelity exists, a greater opportunity for infidelity translates, on average, to a greater probability of infidelity. Although not all infidelity generates sperm competition, the occurrence of sperm competition depends, in large part, on female infidelity (Baker & Bellis, 1995; Smith, 1984).

In summary, men mated to women who are physically attractive, high in Surgency, and high in Openness to Experience may face a high recurrent risk of sperm competition. Ancestral men mated to such women would have reaped reproductive benefits if they were able to prevent or correct their partner's sexual infidelity.

MATE-RETENTION TACTICS

One solution to the adaptive problem of sperm competition is for men to prevent a partner from being sexually unfaithful (Buss, 1988; Buss & Shackelford, 1997; Flinn, 1988; Shackelford, 2003). Buss (1988) identified 19 tactics that people use to guard or to retain their mates, ranging from vigilance over a partner's whereabouts to violence against rivals. Men and women, for example, give to their partners ornaments such as promise, engagement, or wedding rings that signal to others involvement in a romantic relationship. Also, men and women, but particularly men, attempt to prevent partner infidelity by concealing their partner (e.g. refusing to introduce a partner to same-sex friends). Because only women of reproductive age are able to cuckold their partners, Buss and Shackelford (1997) predicted and documented that male mate-retention tactics are more frequent and more intense with partners of high reproductive value, as indicated by the woman's youth and physical attractiveness. Gangestad, Thornhill, and Garver (2002) demonstrated that men increase the frequency and intensity of their mate-retention efforts during the follicular (high-conception-risk) phase of their partner's menstrual cycle.

Because men adjust their mate-retention efforts according to their partner's reproductive value and fertility, perhaps male mate-retention tactics also are sensitive to the recurrent risk of sperm competition. This leads to the first hypothesis, tested in a recent study by Goetz *et al.* (2005):

> **Hypothesis 1:** Recurrent risk of sperm competition will be related positively to the use of mate-retention tactics by men.

SEMEN-DISPLACING BEHAVIORS

Because tactics to prevent a female partner's infidelity are sometimes unsuccessful, male psychology may include mechanisms designed to "correct" a female partner's infidelity (Shackelford, 2003). Inspired by Baker and Bellis's (1993a) demonstration of male *physiological* adaptations to sperm competition, Shackelford *et al.* (2002) documented that human male psychology may include *psychological* adaptations to decrease the likelihood that a rival male's sperm will fertilize a female partner's eggs. For example, men who spent a greater (relative to men who spent a lesser) proportion of time apart from their partner since the couple's last copulation rated their partners as more attractive and reported greater interest in copulating with their partners. As the proportion of time spent apart increases, so too does the likelihood of female infidelity (Baker & Bellis, 1995). Perceiving his partner as more attractive and having an increased interest in copulating with his partner may motivate a man to copulate with his partner as soon as possible, placing his sperm in competition with any rival sperm that may be present in his partner's reproductive tract.

Sperm competition also may have fashioned a psychology that generated specific corrective behaviors designed to increase the likelihood that a man's ejaculate would out-compete rival sperm. This could have been achieved by removing rival semen that was present in his partner's reproductive tract. There is both theory (Baker & Bellis, 1995; Smith, 1984) and empirical evidence (Gallup *et al.*, 2003) suggesting that the human penis may have evolved as a semen-displacement device. Using artificial genitals and simulated semen, Gallup *et al.* (2003) empirically tested the hypothesis that the human penis is designed to displace semen deposited by other men in the reproductive tract of a woman. The results indicated that artificial phalluses with glans and coronal ridge that approximated a real human penis displaced significantly more simulated semen (91%) than did a phallus that did not have a glans and coronal ridge (35%). When the penis is inserted into the vagina, the frenulum of the coronal ridge makes possible semen displacement by allowing semen to flow back under the penis alongside the frenulum and collect on the anterior of the shaft behind the coronal ridge (Gallup et al., 2003). Displacement of simulated semen only occurred, however, when a phallus was inserted at least 75% of its length (which approximates the length of the average human penis) into the artificial vagina.

That the penis must reach an adequate depth before semen is displaced suggests that successfully displacing rival semen requires specific copulatory behaviors. Following allegations of female infidelity or separation from their partners (contexts in which the likelihood of rival semen being present in the reproductive tract is relatively greater), both sexes report that men thrusted deeper and more quickly at the couple's next copulation (Gallup *et al.*, 2003). In addition to thrusting deeply and quickly, other copulatory behaviors may be efficient semen-displacing behaviors. Men who thrusted for a longer time during sexual intercourse, for example, may have been able to displace more semen, thereby "correcting" a woman's recent sexual infidelity. Although in one previous study an extended duration of sexual intercourse did not reliably follow allegations of female infidelity or separation from partners (Gallup *et al.*, 2003), duration of sexual intercourse has been predicted to be a semen-displacing behavior (Gallup & Burch, 2004). The link between the likelihood of female infidelity and semen-displacing behaviors leads to a second hypothesis tested by Goetz *et al.* (2005):

Hypothesis 2: Recurrent risk of sperm competition will be related positively to semen-displacing behaviors.

TACTIC COMPLEMENTARITY

Baker and Bellis (1995) documented that, as female age increases (and reproductive value decreases), the rate of in-pair copulation decreases. High in-pair copulatory frequency has been proposed as a "corrective" measure in the context of sperm competition, because the relative abundance of sperm from the primary male would outnumber rival male sperm due to differential insemination frequency (Parker, 1984). Although the corrective mechanism of frequent in-pair copulations is different than the proposed corrective mechanism of semen-displacing behavior (i.e. outnumbering rival sperm versus displacing rival sperm), both tactics can produce the same result: decreasing the likelihood of cuckoldry. Buss and Shackelford (1997) documented that, as a woman's age increases, her partner's mate-retention effort decreases. The findings from Baker and Bellis (1995) and Buss and Shackelford (1997) suggest that men mated to reproductively valuable women use preventative and corrective tactics simultaneously to solve the adaptive problem of sperm competition.

There is corroborative, comparative evidence that several bird species that practice a socially monogamous mating system similar to humans use mate retention and frequent in-pair copulation as complementary anti-cuckoldry tactics (see, e.g. Dickinson & Leonard, 1996; Møller, 1987). This leads to a third hypothesis tested by Goetz *et al.* (2005):

> **Hypothesis 3:** Male mate-retention tactics and semen-displacing behaviors are complementary tactics designed to solve the adaptive problem of sperm competition. The use of mate-retention tactics therefore will be related positively to semen-displacing behaviors.

In the remainder of this chapter, we highlight the conduct and results of Goetz *et al.* (2005).

Methods

PARTICIPANTS

Three hundred and five men in a committed, sexual relationship with a woman participated in the study. Participants were drawn from universities and surrounding communities. The mean age of the participants was 25.8 years (SD = 8.6 years), and the mean age of the participants' partners was 24.6 years (SD = 8.1 years). Goetz *et al.* (2005) included in the analyses data provided by men who were currently in a relationship that had lasted at least 1 year. This minimum criterion ensures that all participants were involved in a committed, long-term relationship. The mean length of the relationship in which they were involved currently was 52.6 months (SD = 67.9 months).

MATERIALS

Participants completed a survey that included several sections. The first section requested demographic information, such as the participant's age and his partner's age. The second section asked four questions to assess partner attractiveness: How physically attractive do you think your partner is?, How physically attractive do other men think she is?, How sexually attractive do you think she is?, and How sexually attractive do other men think she is? We recorded responses using a Likert-type scale ranging from 0 (Not at all) to 9 (Extremely). The third section asked the participant about his copulatory behaviors with his current partner, compared to his typical copulatory behaviors in the past. To standardize the time frame of reports, Goetz *et al.* (2005) asked the participant about the most recent sexual encounter with his current partner.

The researchers assessed several copulatory behaviors, including number of thrusts, deepest thrust, depth of thrusts on average, and duration of sexual intercourse. They recorded responses using a Likert-type scale ranging from 0 (Lesser/Shorter/Fewer) to 9 (Greater/Longer/More). The fourth section asked how often the participant performed 104 mate-retention acts (from Buss, 1988) in the last month, ranging from 0 (Never) to 3 (Often). Example mate-retention acts include, "Refused to introduce my partner to same-sex friends," "Read my partner's personal mail," and "Bought my partner a small gift." The final section requested a participant's assessments of his partner's standings on the major dimensions of personality. This personality inventory included 40 bipolar items designed to assess standings on five major dimensions of personality (Botwin, Buss, & Shackelford, 1997; example item anchors are in parentheses): Surgency (dominant – submissive), Agreeableness (flexible – stubborn), Conscientiousness (careless – careful), Emotional Stability (secure – insecure), and Openness to Experience (uncultured – cultured). For each item, participants circled a number between 1 and 7 inclusive that described their partner "generally." This measure provides reliable and valid assessments of the five major dimensions of personality and, in addition, previous work indicates substantial agreement between ratings about a target person provided by (1) that target person and (2) the target person's long-term romantic partner (see Botwin *et al.*, 1997).

PROCEDURE

Three criteria must have been met to qualify for participation: the prospective participant must be (1) male, (2) at least 18 years of age, and (3) currently involved in a committed, romantic, sexual relationship with a woman. Prospective participants were aware of these participation criteria prior to appearing at a scheduled time and location. Upon the prospective participant's arrival at the schedule time and location, the research assistant confirmed that the prospective participant met the three participation criteria.

If the participation criteria were met, the research assistant handed the participant a consent form, the survey, and a security envelope. The participant was instructed not to seal the consent form inside the envelope to maintain anonymity. The research assistant explained to the participant the purpose of the study, answered any questions, and thanked the participant for his participation.

Results

Goetz *et al.* (2005) created several composite variables prior to analyses. Partner's attractiveness ($\alpha = 0.86$) is the mean of four variables: participants' rating of partner's (1) physical attractiveness and (2) sexual attractiveness,

Table 9.1. *Correlations between mate-retention tactics and recurrent risk of sperm competition, and between mate-retention tactics and semen-displacing behaviors.*

Mate retention tactic (α) [number of acts]	Recurrent risk of sperm competition	Semen-displacing behaviors
Vigilance (0.82) [9 acts]	0.06	0.09
Concealment of mate (0.65) [4 acts]	−0.13*	0.07
Monopolization of mate (0.75) [5 acts]	−0.03	0.04
Jealousy inducement (0.72) [4 acts]	−0.09	0.01
Punishment of threat to infidelity (0.82) [7 acts]	−0.01	0.09
Emotional manipulation (0.81) [8 acts]	−0.03	0.05
Commitment manipulation (0.50) [3 acts]	0.13*	0.24‡
Derogation of competitors (0.77) [7 acts]	−0.08	0.01
Resource display (0.86) [6 acts]	0.28‡	0.24‡
Sexual inducement (0.62) [5 acts]	0.14*	0.16†
Appearance enhancement (0.77) [5 acts]	0.22‡	0.13*
Love and care (0.67) [5 acts]	0.31‡	0.17‡
Submission and debasement (0.68) [5 acts]	0.05	0.09
Verbal possession signals (0.65) [5 acts]	0.20‡	0.18†
Physical possession signals (0.72) [5 acts]	0.27‡	0.13*
Possessive ornamentation (0.66) [5 acts]	0.16†	0.16†
Derogation of mate (0.70) [5 acts]	−0.24‡	0.01
Intrasexual threats (0.84) [6 acts]	0.05	0.12*
Violence against rivals (0.82) [5 acts]	−0.05	0.11

Note: From Goetz *et al.* (2005); $n = 305$. $\alpha = \alpha$ reliability.
*$P < 0.05$; †$P < 0.01$; ‡$P < 0.001$ (two-tailed).

and participants' rating of other men's assessments of partner's (3) physical attractiveness and (4) sexual attractiveness. Semen-displacing behavior ($\alpha = 0.81$) is the mean of four variables: (1) number of thrusts, (2) deepest thrust, (3) depth of thrusts on average, and (4) duration of sexual intercourse.

Goetz *et al.* (2005) calculated scores for 19 mate-retention tactics using responses to the 104 mate-retention acts, following Buss (1988; and see Buss & Shackelford, 1997). The researchers calculated scores for participants' partners on the five major dimensions of personality using responses to the Botwin *et al.* (1997) 40-item instrument. α Reliabilities for the 19 mate-retention tactics varied in this study from $\alpha = 0.50$ to 0.86 (see Table 9.1). For use in several statistical analyses, the researchers created a composite mate-retention variable ($\alpha = 0.97$) by averaging responses on the 19 mate-retention tactics. α Reliabilities for the target partner personality dimensions were: Surgency, $\alpha = 0.62$ and Openness to Experience, $\alpha = 0.63$. The researchers created a composite recurrent risk of sperm competition variable ($\alpha = 0.66$) by averaging scores on partner's attractiveness, partner's Surgency, and partner's Openness to Experience. The researchers scaled partner's attractiveness differently from the personality measures, so they standardized the three variables prior to averaging.

Hypothesis 1 stated that recurrent risk of sperm competition will be related positively to the use of mate-retention tactics by men. Consistent with this hypothesis, recurrent risk of sperm competition correlated positively with the use of mate-retention tactics; $r (303) = 0.14$ ($P < 0.05$). Men mated to women who place them at a higher recurrent risk of sperm competition reported more frequent use of mate-retention tactics. Although the focus of Hypothesis 1 was the relationship between recurrent risk of sperm competition and the broad use of mate-retention tactics, Goetz et $al.$ (2005) also investigated the relationship between recurrent risk of sperm competition and use of each of the 19 mate-retention tactics (see Table 9.1) reported by Buss (1988) and Buss and Shackelford (1997). Eight of the 19 tactics showed significant positive correlations with recurrent risk of sperm competition (commitment manipulation, resource display, sexual inducement, appearance enhancement, love and care, verbal possession signals, physical possession signals, and possessive ornamentation), and two of the 19 tactics showed significant negative correlations with recurrent risk of sperm competition (concealment of mate and derogation of mate).

Hypothesis 2 stated that recurrent risk of sperm competition will be related positively to semen-displacing behaviors. Consistent with this hypothesis, recurrent risk of sperm competition correlated positively with semen-displacing behaviors; $r (303) = 0.33$ ($P < 0.001$). Men mated to women who place them at a higher recurrent risk of sperm competition reported performing more copulatory behaviors arguably designed to displace rival semen. For reportorial completeness, Goetz et $al.$ (2005) also investigated the relationship between recurrent risk of sperm competition and use of each of the four individual semen-displacing behaviors (see Table 9.2). All four correlations were significant and positive, ranging from $r (303) = 0.21$ to 0.32 (all P values < 0.001).

Hypothesis 3 stated that the use of mate-retention tactics will be related positively to semen-displacing behaviors. Consistent with this hypothesis, the use of mate-retention tactics correlated positively with semen-displacing behaviors; $r (303) = 0.19$ ($P < 0.01$). This positive correlation suggests that men used mate-retention tactics and semen-displacing behaviors simultaneously. Goetz et $al.$ (2005) also investigated the relationship between semen-displacing behaviors and each of the 19 mate-retention tactics reported by Buss (1988) and Buss and Shackelford (1997; see Table 9.1). Nine of the 19 tactics were positively and significantly correlated with semen-displacing behaviors (commitment manipulation, resource display, sexual inducement, appearance enhancement,

Table 9.2. *Correlations between semen-displacing behaviors and recurrent risk of sperm competition.*

Semen-displacing behaviors	Recurrent risk of sperm competition
Number of thrusts	0.22[‡]
Deepest thrust	0.32[‡]
Depth of thrusts, on average	0.31[‡]
Duration of sexual intercourse	0.21[‡]

Note: From Goetz *et al.* (2005); $n = 305$. Each semen-displacing behavior is a relative measure; the participant compared his copulatory behaviors with his current partner to his typical copulatory behaviors in the past.
[‡] $P < 0.001$ (two-tailed).

love and care, verbal possession signals, physical possession signals, possessive ornamentation, and intrasexual threats).

Discussion

Shackelford (2003) proposed three separate adaptive problems associated with sperm competition in human males: preventing a female partner's sexual infidelity, correcting a female partner's sexual infidelity, and anticipating a female partner's sexual infidelity. The research by Goetz *et al.* (2005) represents a preliminary investigation of how men might solve the adaptive problems of preventing and correcting a partner's infidelity. More specifically, this study tested the hypotheses that men mated to women that recurrently place them at a high risk of sperm competition may solve this adaptive problem through preventative and corrective measures, and that the preventative and corrective measures are complementary, working together to solve the adaptive problem of sperm competition. Behaviors that functioned to prevent and correct a female partner's sexual infidelity may have been selected for in a socially monogamous species such as humans (Baker & Bellis, 1995; Shackelford, 2003; Smith, 1984). The results suggest that men perform both preventative and corrective behaviors when under a high recurrent risk of sperm competition.

Women who are physically attractive and have personality characteristics that attract rival men are more often given the opportunity to be sexually unfaithful and may be more likely to commit sexual infidelity, thereby placing their partners at a higher recurrent risk of sperm competition. The female partner's physical attractiveness, Surgency, and Openness to Experience index a man's recurrent risk of sperm competition. Men mated to such women may solve the adaptive problem of sperm competition through preventative and corrective measures. Goetz *et al.* (2005) operationalized preventative measures as the use of mate-retention tactics (Buss, 1988), and we operationalized corrective measures as male copulatory behaviors consisting of relatively deep, numerous thrusts for a prolonged period of time – behaviors that may be effective at displacing rival semen present in the reproductive tract of a woman (Gallup *et al.*, 2003).

The results supported all three hypotheses. Men mated to women who have traits linked to a higher probability of sexual infidelity more frequently use mate-retention tactics apparently designed to thwart potential infidelity. Men mated to women who have traits linked to a higher probability of sexual infidelity also are more likely to perform copulatory behaviors arguably designed to displace rival male semen present in the woman's reproductive tract. Finally, the results indicate that male mate-retention tactics and semen-displacing behaviors are complementary tactics used simultaneously to reduce sperm-competition risk and the consequences of sperm competition (e.g. cuckoldry).

The majority of copulations in humans are in-pair copulations (Baker & Bellis, 1995) and, therefore, semen-displacing behaviors performed by the primary male usually will displace his own semen. The consequences of such an effect might be minimized, however, if the temporal spacing between successive in-pair copulations is greater than the spacing between copulations involving different men. The refractory period may have been designed for this purpose (Gallup & Burch, 2004). Moreover, Gallup and Burch (2004) have suggested that penile hypersensitivity and loss of an erection (two events that follow ejaculation) may function to reduce the occurrence of self-semen displacement.

Although the sooner a man performs semen-displacing behaviors the more likely he is to displace recently deposited rival semen, it is not clear, at present, how long the "corrective window" may be. Because the human penis cannot enter the cervix to displace semen, the corrective window should be proportional to the time that sperm remain in the vagina. Although some sperm have been found in the cervix and oviducts within minutes of coitus, they are often dead (Johnson & Everitt, 1995; Porter & Flinn, 1977). Evidence from reproductive biology suggests that human sperm can be stored for varying time periods within the vaginal cavity (see, e.g. Baker & Bellis, 1995; Johnson & Everitt, 1995; Morris, 1977; Smith, 1984). One way to determine the length of the corrective window is to establish the proportion of time that sperm remain in the vagina. Another way to determine the length of the corrective window may be to correlate semen-displacing behaviors as a function of time since last sexual intercourse. The length of the corrective window will be revealed as semen-displacing behaviors begin to decrease as the time since last sexual intercourse increases.

Most of the relationships assessed by Goetz *et al.* (2005) between recurrent risk of sperm competition and mate-retention tactics are positive, indicating that men mated to women who place them at high recurrent risk of sperm competition more frequently perform mate-retention behaviors. Recurrent risk of sperm competition correlates positively with the mate-retention tactic sexual inducement, for example. Sexual inducement includes the following items: "He gave in to her sexual requests," "He performed sexual favors to keep his partner around," "He had a physical relationship with her to deepen their bond," and "He gave in to sexual pressure to keep her." Although the phrasing of these items suggests that sexual inducement is used more often by women, Buss (1988) found a sex difference in the opposite direction (i.e. men reported using sexual inducement more than women). From the perspective of sperm-competition theory (Parker, 1970), the sexual inducement tactic can be interpreted as a "corrective" tactic designed to place a man's sperm in competition with any rival sperm that may be present in his partner's reproductive tract.

Limiting factors such as time, energy, and resources do not allow men to perform every mate-retention tactic all the time, and some tactics may be avoided deliberately. Many of the non-significant correlations between recurrent risk of sperm competition and the mate-retention tactics involve "negative" tactics (i.e. tactics likely to be perceived negatively by one's partner), such as vigilance, monopolization of time, and emotional manipulation. Performing negative mate-retention tactics is likely to excite conflict in a relationship. Goetz *et al.* (2005) speculated that, to avoid such conflict, men mated to women who are more likely to commit sexual infidelity do not attempt to retain their mates using these negative tactics.

Indeed, recurrent risk of sperm competition correlated negatively and significantly with the mate-retention tactics derogation of mate and concealment of mate. Derogation of mate includes items such as, "He told other guys she was not a nice person," "He told other guys she was stupid," and "He told others she was a pain." Perhaps men mated to women who are more likely to commit sexual infidelity do not attempt to retain their mates by derogating their mates to competitors because this derogation may signal to competitors impending relationship dissolution, prompting competitors to pursue the woman. Directing rival men to a mate who is likely to commit sexual infidelity is unlikely to have been a reproductively profitable strategy.

Examining the relationship between recurrent risk of sperm competition and each of the four semen-displacing behaviors reveals that men mated to women who place them at high recurrent risk of sperm competition report that during sexual intercourse with their partner (1) they thrust more frequently, (2) their deepest thrust is more deep, (3) the depth of their thrusts is much deeper on average, and (4) the duration of sexual intercourse is much longer. That men report that their deepest thrust is more deep than usual and that the depth of their thrusts is much deeper on average corroborates Gallup et al.'s (2003) demonstration that the penis must reach an adequate depth to displace semen successfully. The other copulatory behaviors – number of thrusts and duration of sexual intercourse – also may contribute to semen displacement. Thrusting more frequently during intercourse and a longer duration of intercourse may afford a man greater opportunity to displace rival semen. Although not found in Gallup et al.'s (2003) work, duration of sexual intercourse was predicted to be a semen-displacing behavior by Gallup and Burch (2004).

All of the relationships assessed by Goetz et al. (2005) between semen-displacing behaviors and mate-retention tactics are positive and several achieve statistical significance, suggesting that the two sets of behaviors are used simultaneously to solve the adaptive problem of sperm competition. This concurs with cross-species evidence that several bird species use preventative and corrective tactics as complementary anti-cuckoldry tactics (see, e.g. Dickinson & Leonard, 1996; Møller, 1987).

Semen-displacing behaviors correlate positively with the mate-retention tactic sexual inducement, for example. When men are giving in to their partners' sexual requests, performing sexual favors to keep their partners around, having a physical relationship with their partners to deepen their bond, and giving in to sexual pressure to keep their partner, they are performing semen-displacing behaviors during these acts. This relationship supports the interpretation of sexual inducement as a corrective tactic designed to (a) place a man's sperm in competition with any rival sperm that may be present in his partner's reproductive tract, or (b) displace semen deposited by rival males.

An alternative explanation for the occurrence of semen-displacing behaviors in men who are mated to women who place them at a high recurrent risk of sperm competition might be that such men thrust more deeply and frequently, for example during sexual intercourse, because they are extremely sexually aroused as a result of their partners' physical attractiveness (a key component of high recurrent risk of sperm competition). Proponents of the "heightened sexual arousal" hypothesis must answer why men are more sexually aroused by physically attractive partners, however. Shackelford et al. (2002) argued that perceiving his partner as more attractive and becoming more sexually aroused may motivate a man to copulate with his partner as soon as possible, placing his sperm in competition with any rival sperm that may be present in his partner's reproductive tract. So the heightened sexual arousal hypothesis is not an alternate hypothesis.

Another alternative explanation for the occurrence of semen-displacing behaviors in men who are mated to women who place them at a high recurrent risk of sperm competition might be that women who are attractive, sexy, dominant, sociable, curious, creative, and knowledgeable are simply more open to intense, varied, and prolonged copulatory behavior with their partners. To address this possibility, Goetz et al. (2005) assessed the woman's sexual openness by asking the participant, "In general, how open is your partner about sexual matters?" Goetz et al. identified no significant correlations between sexual openness and physical attractiveness, Surgency, and Openness to Experience. Thus, attractive, sexy, dominant, sociable, curious, creative, knowledgeable women are not simply more open to their partners' intense, varied, and prolonged copulatory behaviors.

LIMITATIONS AND FUTURE RESEARCH DIRECTIONS

This research by Goetz et al. (2005) has several limitations. One limitation of the study is in its design. Goetz et al. present correlational analyses that prevent strong statements about causal relationships. Goetz et al. speculate that mateship to women who have traits linked to a higher probability of sexual infidelity *causes* men to perform mate-retention tactics and semen-displacing behaviors. The data are consistent with this interpretation, but we cannot yet rule out an alternative, reverse causal relationship – that performing preventative and corrective measures causes men to select as mates women who have traits linked to a higher probability of sexual infidelity. A methodology that includes repeated assessments of the key variables over time, such as a daily diary study, would allow for the identification of causal relationships.

An arguable design limitation of this research is the use of men's reports of their female partners' attractiveness and personality standings. Perhaps the results would have turned out differently if Goetz et al. (2005) had collected independent ratings on these key variables. This is not likely, however, because previous research indicates that men's ratings of their female partners' attractiveness and personality standings correlate positively with independent ratings (and with women's self-reports) on these variables (see, e.g. Botwin et al., 1997; Buss & Shackelford, 1997). Furthermore, it is the male's perception of his partner's physical attractiveness and personality that are important. The male's perception provides the input that is transformed through psychological processes (i.e. decision rules) into behavioral output (e.g. perform semen-displacing behaviors or do not perform semen-displacing behaviors).

There may be concern about the validity of the self-reported assessments of specific copulatory behaviors. The validity of this self-report methodology has been established in several different ways. Masters and Johnson (1966) provided some of the first evidence for the substantial agreement between self-reports of specific copulatory behaviors and these same behaviors as observed directly and recorded by the researchers. In several dozen studies with several hundred participants, Masters and Johnson (1966) documented that people can accurately report the occurrence of specific copulatory behaviors. For the majority of specific copulatory behaviors, recollected reports of these behaviors are in substantial agreement with the actual behaviors observed or recorded by the researchers. As is true for most recollections, this agreement between self-reports and actual behavior is greater the closer in time the recollections are to the actual behaviors. This agreement is substantial, however, even after a period of several weeks or months (see research reviewed in Allgeier & Allgeier, 2000; Crooks & Baur, 2002; Hyde & DeLamater, 2003; and Masters, Johnson, & Kolodny, 1997). Goetz *et al.* (2005) assessed self-reported copulatory behaviors that occurred at the most recent sexual encounter with the participant's regular partner. In this relatively young sample, it is typical that most individuals have sexual intercourse with their partners two to three times per week (see Baker & Bellis, 1995; Smith, 1984). Hence, the last sexual encounter is not likely to have occurred more than 2 or 3 days prior to the survey administration. This is not a long period over which to provide recollections and, therefore, the self-reported assessments of specific copulatory behaviors should not prove to be problematic.

There are several additional directions for future work. Future research should examine semen-displacing behaviors following an actual female sexual infidelity. Goetz *et al.* (2005) examined the relationships between the *likelihood* of female sexual infidelity (as assessed by her physical attractiveness and personality traits) and male mate-retention and semen-displacing behaviors. More intense mate-retention and semen-displacing behaviors may be detectable under conditions of suspected or confirmed sexual infidelity.

Future research also might attempt to establish the woman's role when men employ these preventative and corrective measures, particularly because the interests of men and women often are in conflict. If she is seeking genetic benefits in her extra-pair copulations (Cashdan, 1996; Gangestad & Simpson, 2000; Scheib, 2001), for example, a woman who was inseminated recently by an extra-pair partner may resist or find unpleasant her regular partner's "corrective" semen-displacing behaviors. In addition, we might hypothesize that she might counter her partners' semen-displacing behaviors by failing to achieve a sperm-retaining orgasm with her regular partner (Baker & Bellis, 1993b, 1995).

In conclusion, a wide variety of human male psychological adaptations may have been designed by selection due to sperm competition. To prevent sperm competition, men perform mate-retention tactics apparently designed to reduce the likelihood of female sexual infidelity. Because preventative tactics are not fail-safe, however, men perform specific copulatory behaviors apparently designed to correct female sexual infidelity by displacing rival semen that may be present in the woman's reproductive tract. These tactics may accompany physiological adaptations (Baker & Bellis, 1993a) and other psychological adaptations (Shackelford *et al.*, 2002) to solve the adaptive problem of human sperm competition.

References

Allgeier, E. R. and Allgeier, A. R. (2000). *Sexual Interactions*, 5th edn. Boston: Houghton Mifflin.

Baker, R. R. and Bellis, M. A. (1993a). Human sperm competition: ejaculate adjustment by males and the function of masturbation. *Animal Behaviour*, **46**, 861–85.

Baker, R. R. and Bellis, M. A. (1993b). Human sperm competition: ejaculate manipulation by females and a function for the female orgasm. *Animal Behaviour*, **46**, 887–909.

Baker, R. R. and Bellis, M. A. (1995). *Human Sperm Competition*. London: Chapman & Hall.

Baur, B. (1998). Sperm competition in molluscs. In T. R. Birkhead and A. P. Møller, eds., *Sperm Competition and Sexual Selection*. San Diego: Academic Press, pp. 255–305.

Birkhead, T. R. and Møller, A. P. (1992). *Sperm Competition in Birds*. London: Academic Press.

Birkhead, T. R. and Møller, A. P. (eds.). (1998). *Sperm Competition and Sexual Selection*. San Diego: Academic Press.

Botwin, M. D., Buss, D. M., and Shackelford, T. K. (1997). Personality and mate preferences: five factors in mate selection and marital satisfaction. *Journal of Personality*, **65**, 107–36.

Buss, D. M. (1988). From vigilance to violence: tactics of mate retention in American undergraduates. *Ethology and Sociobiology*, **9**, 291–317.

Buss, D. M. (1989). Sex differences in human mate preferences: evolutionary hypotheses tested in 37 cultures. *Behavioral and Brain Sciences*, **12**, 1–49.

Buss, D. M. and Schmitt, D. (1993). Sexual strategies theory: an evolutionary perspective on human mating. *Psychological Review*, **100**, 204–32.

Buss, D. M. and Shackelford, T. K. (1997). From vigilance to violence: mate retention tactics in married couples. *Journal of Personality and Social Psychology*, **72**, 346–61.

Buss, D. M., Larsen, R. J., Westen, D., and Semmelroth, J. (1992). Sex differences in jealousy: evolution, physiology and psychology. *Psychological Science*, **3**, 251–5.

Cashdan, E. (1996). Women's mating strategies. *Evolutionary Anthropology*, **5**, 134–43.

Crooks, R. L. and Baur, K. (2002). *Our sexuality* (8th edn). Pacific Grove, CA: Brooks/Cole.

Daly, M., Wilson, M., and Weghorst, J. (1982). Male sexual jealousy. *Ethology and Sociobiology*, **3**, 11–27.

Dickinson, J. L. and Leonard, M. L. (1996). Mate attendance and copulatory behaviour in western bluebirds: evidence of mate guarding. *Animal Behaviour*, **52**, 981–92.

Dijkstra, P. and Buunk, B. P. (2001). Sex differences in the jealousy-evoking nature of a rival's body build. *Evolution and Human Behavior*, **22**, 335–41.

Flinn, M. V. (1988). Mate guarding in a Caribbean village. *Ethology and Sociobiology*, **9**, 1–28.

Gallup G. G. and Burch, R. L. (2004). Semen displacement as a sperm competition strategy in humans. *Evolutionary Psychology*, **2**, 12–23.

Gallup G. G., Burch, R. L., Zappieri, M. L., *et al.* (2003). The human penis as a semen displacement device. *Evolution and Human Behavior*, **24**, 277–89.

Gangestad, S. W. and Simpson, J. A. (1990). Toward an evolutionary history of female sociosexual variation. *Journal of Personality*, **58**, 69–96.

Gangestad, S. W. and Simpson, J. A. (2000). The evolution of human mating: trade-offs and strategic pluralism. *Behavior and Brain Sciences*, **23**, 573–87.

Gangestad, S. W., Thornhill, R., and Garver, C. E. (2002). Changes in women's sexual interests and their partner's mate-retention tactics across the menstrual cycle: evidence for shifting conflicts of interest. *Proceedings of the Royal Society of London B*, **269**, 975–82.

Goetz, A. T., Shackelford, T. K., Weekes-Shackelford, V. A., *et al.* (2005). Mate retention, semen displacement, and human sperm competition: a preliminary investigation of tactics to prevent and correct female infidelity. *Personality and Individual Differences*, **38**, 749–63.

Goldberg, L. R. (1982). From ace to zombie: some explorations in the language of personality. In C. D. Spielberg and J. N. Butcher, eds., *Advances in Personality Assessment*. Hillsdale, NJ: Erlbaum, vol. 1, pp. 203–34.

Harris, C. R. (2003). A review of sex differences in sexual jealousy, including self-report data, psychophysiological responses, interpersonal violence, and morbid jealousy. *Personality and Social Psychology Review*, **7**, 102–28.

Hughes, S. M. and Gallup, G. G. (2003). Sex differences in morphological predictors of sexual behavior: shoulder to hip and waist to hip ratios. *Evolution and Human Behavior*, **24**, 173–8.

Hyde, J. S. and DeLamater, J. (2003). *Understanding Human Sexuality*, 8th edn. Boston: McGraw-Hill.

Johnson, M. H. and Everitt, B. J. (1995). *Essential Reproduction*, 4th edn. Oxford: Blackwell Science.

Li, N. P., Bailey, J. M., Kenrick, D. T., and Linsenmeier, J. A. W. (2002). The necessities and luxuries of mate preferences: testing the tradeoffs. *Journal of Personality and Social Psychology*, **82**, 947–56.

Masters, W. H. and Johnson, V. E. (1966). *Human Sexual Response*. Boston: Little, Brown, & Co.

Masters, W. H., Johnson, V. E., & Kolodny, R. C. (1997). *Human Sexuality*. New York: Harper Collins.

Møller, A. P. (1987). Mate guarding in the swallow *Hirundo rustica*. *Behavioral Ecology & Sociobiology*, **21**, 119–23.

Morris, J. M. (1977). The morning-after pill: a report on postcoital contraception and interception. In R. O. Greep and M. A. Koblinsky, eds., *Frontiers in Reproductive and Fertility Control*. Cambridge, MA: MIT Press, pp. 203–8.

Norman, W. T. (1963). Toward an adequate taxonomy of personality attributes: replicated factor structure in peer nominations and personality ratings. *Journal of Personality and Social Psychology*, **66**, 574–83.

Parker, G. A. (1970). Sperm competition and its evolutionary consequences in the insects. *Biological Review*, **45**, 525–67.

Parker, G. A. (1984). Sperm competition and the evolution of animal mating strategies. In R. L. Smith, ed., *Sperm Competition and the Evolution of Animal Mating Systems*. London: Academic Press, pp. 1–60.

Porter, D. G. and Flinn, C. A. (1977). The biology of the uterus. In R. O. Greep & M. A. Koblinsky, eds., *Frontiers in Reproductive and Fertility Control*. Cambridge, MA: MIT Press, pp. 146–56.

Regan, P. C., Levin, L., Sprecher, S., Christopher, F. S., and Cate, R. (2000). Partner preferences: what characteristics do men and women desire in their short-term sexual and long-term partners? *Journal of Psychology & Human Sexuality*, **12**, 1–21.

Scheib, J. E. (2001) Context-specific mate choice criteria: women's trade-offs in the contexts of long-term and extra-pair mateships. *Personal Relationships*, **8**, 371–89.

Schmitt, D. P. and Buss, D. M. (2001). Human mate poaching: tactics and temptations for infiltrating existing mateships. *Journal of Personality and Social Psychology*, **80**, 894–917.

Shackelford, T. K. (2003). Preventing, correcting, and anticipating female infidelity: three adaptive problems of sperm competition. *Evolution and Cognition*, **9**, 90–6.

Shackelford, T. K., LeBlanc, G. J., Weekes-Shackelford, V. A., *et al.* (2002). Psychological adaptation to human sperm competition. *Evolution and Human Behavior*, **23**, 123–38.

Shackelford, T. K., Goetz, A. T., LaMunyon, C. W., *et al.* (2004). Sex differences in sexual psychology produce sex similar preferences for a short-term mate. *Archives of Sexual Behavior*, **33**, 405–12.

Simmons, L. W. (2001). *Sperm Competition and its Evolutionary Consequences in the Insects*. Princeton, NJ: Princeton University Press.

Singh, D. (1993). Adaptive significance of waist-to-hip ratio and female physical attractiveness. *Journal of Personality and Social Psychology*, **65**, 293–307.

Smith, R. L. (1984). Human sperm competition. In R. L. Smith, ed., *Sperm Competition and the Evolution of Animal Mating Systems*. New York: Academic Press, pp. 601–60.

Streeter, S. A. and McBurney, D. H. (2003). Waist-hip ratio and attractiveness: new evidence and a critique of "a critical test". *Evolution and Human Behavior*, **24**, 88–98.

Symons, D. (1979). *The Evolution of Human Sexuality*. New York: Oxford University Press.

White, G. L. and Mullen, P. E. (1989). *Jealousy*. New York: Guilford Press.

Wyckoff, M., Wang, W., and Wu, C. (2000). Rapid evolution of male reproductive genes in the descent of man. *Nature*, **403**, 304–8.

10

Preeclampsia and other pregnancy complications as an adaptive response to unfamiliar semen

JENNIFER A. DAVIS AND GORDON G. GALLUP JR.
State University of New York at Albany

Preeclampsia

Preeclampsia is a leading cause of prenatal infant mortality (Mac Gillivray, 1983; Robillard, Dekker, & Hulsey, 2002). Preeclampsia occurs as a consequence of abnormal invasion by the trophoblast in the uterine spiral arteries and endothelial cell dysfunction (Friedman, 1993), and as a consequence the fetus may not receive adequate nutrition resulting in growth retardation. Whereas all mammalian embryos undergo implantation shortly after conception, humans are the only mammalian species known to undergo a second phase of deep trophoblastic implantation at the end of the first trimester (Robillard *et al.*, 2003). In normal development, this second stage of implantation provides for the modification of spiral arteries that result in an increase in the blood flow to the placenta. Preeclampsia is believed to be the result of a failure to achieve or to complete this second implantation phase (Robillard *et al.*, 2003). It is clinically diagnosed by maternal hypertension and proteinuria. The hypertension results from cytotrophic factors that are released by the fetus and serve to increase the amount of blood flowing to the placenta (Haig, 1993).

It has been theorized that the origins of preeclampsia in humans are linked to the increase in cranial capacity associated with the genus *Homo* (Robillard *et al.*, 2003). The greater nutritional needs of the developing brain in the human fetus, compared to the more modest needs of developing brains in species with lower cranial capacities, has been hypothesized to explain the second wave of implantation characteristic of humans. To support normal brain development, the human fetal brain requires approximately 60% of the total nutritional supplies of the developing fetus at the end of the third trimester, whereas in other mammalian species developing fetal brains require only about 20% of the total nutritional needs of the fetus toward the end of fetal development (Cunnane, Harbige, & Crawford, 1993; Martin, 1996). Thus, the second wave of implantation may be an important accommodation to the evolution of relative large brains in humans.

In this chapter we develop the thesis that this second phase of implantation in humans also may have evolved to enable females to unwittingly terminate pregnancies that were not in their long-term reproductive interests.

The role of unfamiliar semen

Preeclampsia involves a host of immunologic and genetic factors (Pridjian & Puschett, 2002; Scott & Beer, 1976), but because the underlying cause remains unknown it has been referred to as "mysterious" (Takakuwa *et al.*, 1999) and the "disease of theories" (Pipkin & Rubin, 1994).

We begin our analysis by exploring the potential significance and ostensible impact of unfamiliar semen on preeclamptic pregnancies. Guided by an evolutionary perspective, we contend that unfamiliar semen may be a biological correlate of paternal investment. According to our hypothesis, pregnancies and children that result from unfamiliar semen have a lower probability of receiving sufficient paternal investment than do pregnancies and children that result from familiar semen. We theorize that preeclampsia is a biological mechanism that evolved to terminate maternal investment under circumstances in which the likelihood of investment by the sire is doubtful.

The importance of paternal investment

Lack of paternal investment poses substantial risks to both the mother and her infant. Ancestral women who lacked the care, protection, and provisioning afforded by a committed adult male would have put themselves and their infants at an increased risk of predation, had a more difficult time acquiring necessary resources, and lowered their chances of finding another provisioning male with whom to have children. In modern times, the costs incurred from the lack of paternal investment remain high. When children's mental, physical, and emotional development has been investigated as a function of paternal investment, the presence and involvement of fathers has consistently been shown to have a positive impact on their child's health.

For instance, children who are involved with their fathers tend to be more psychologically well adjusted, do better in school, engage in less anti-social behavior, and have more successful intimate relationships (Amato & Rivera,

1999; Flouri & Buchanan 2002, 2003; Hwang & Lamb, 1997). Father involvement also has been associated with positive intellectual development, social competence, and the ability to empathize (Fagan and Iglesias, 1999; Yongman, Kindlon & Earls 1995). Conversely, the absence of the father is related to poorer scores on measures of cognitive ability and academic achievement, and these children have a heightened risk for delinquency and deviant behavior (Dornbusch *et al.*, 1985; Mulkey, Crain, & Harrington, 1992).

Gaudino, Jenkins, and Rouchat (1999) examined the risk posed to children by the presence or absence of the father's name on the birth certificate. Presumably, acknowledgment of the father's name on the birth certificate would be an indicator of the father being more likely to have a closer relationship with the mother and the infant, and that these fathers are more likely to be in involved in the support and care of their infants. Gaudino *et al.* found that infants without reported fathers' names were 2.3 times more likely to die in the first year of life. Even after controlling for variables such as marital status, maternal variables such as age, race, education, gravidity, and smoking, and factors that may affect infants, such as congenital malformations and birth weight, the absence of the father's name on the birth certificate remained an independent risk factor for infant mortality.

Other studies suggest that paternal involvement reduces the risk of low birth weight and preterm delivery (Mutale *et al.*, 1991; Oakley, 1985; Ramsey, Abell, & Baker, 1986; Turner, Grindstaff, & Phillips, 1990). Low birth weight is an index of overall infant health and is related to infant mortality, health problems in the first few years of life, and developmental problems later in life (Hack, Klein, & Taylor, 1995).

Unmarried women are consistently at greater risk of delivering low-birth-weight babies (Jones & Bond, 1999). Among unmarried women, the extent to which the father is involved also influences birth weight. In a sample of unmarried women, Padilla and Reichman (2001) found that low birth weight varied as a function of the mother's relationship to the father. Mothers who did not live with the infant's father were more likely to deliver low-birth-weight infants. In contrast, mothers who received financial support from the baby's father were at decreased risk of delivering low-birth-weight babies.

Semen familiarity and paternity

In humans, paternal investment has far-reaching implications for both the mother and the infant. One adaptation that may have evolved to solve the ancestral problem posed by both the mother's and offspring's need for a caring, committed father is a maternal mechanism that could distinguish between familiar and unfamiliar semen. Frequent insemination by the same male over an extended period of time would be a relatively good index of a committed pair bond and, therefore, semen familiarity might predict the likelihood of long-term provisioning, protection, and care of the mother and child by the father following parturition. Therefore, we hypothesize that preeclampsia may be a mechanism that evolved to terminate pregnancies that result from exposure to unfamiliar semen.

In support of this hypothesis, exposure to unfamiliar semen (as would be true in cases of barrier contraception, artificial insemination, instances of changed paternity in multiple pregnancies, and a short period of cohabitation before pregnancy) increases the risk of developing preeclampsia (Astin, Scott, & Worley, 1981; Chng, 1982; Dekker, Robillard, & Hulsey, 1998; Feeney & Scott, 1980; Ikedife, 1980; Klonoff-Cohen et al., 1989; Koelman et al., 2000; Marti & Hermann, 1977; Robillard et al., 1993; Robillard, Hulsey, & Alexander, 1980).

For example, Hoy et al. (1999) studied the outcomes of 1552 donor-insemination pregnancies and 7717 normally conceived pregnancies in Australia. They found that the donor-insemination group was more likely to develop preeclampsia than the controls, even after controlling for maternal age and parity. In another study, the incidence of preeclampsia also was found to be more common in donor inseminations, especially in multiparous women (Need et al., 1983). As evidence that these effects are not a consequence of artificial insemination per se, Smith, et al. (1997) report that preeclampsia was more common in women who received donor insemination than in patients who were artificially inseminated with their committed partner's sperm.

The extended presence of semen in the female reproductive tract as a byproduct of repeated insemination has a significant effect on improving rates of both implantation and pregnancy (Bellinge et al., 1986; Marconi et al., 1989). Thaler (1989) suggests that familiar semen may exert a priming effect on pregnancy acceptance. This prediction is supported by clinical trials in which females are given prior exposure to semen before conception is attempted (Bellinge et al., 1986; Marconi et al., 1989).

Semen familiarity also varies as a function of whether couples use barrier methods of contraception. In a study comparing the contraceptive histories of nulliparous females with and without preeclampsia, twice as many of those with preeclampsia used barrier methods of birth control (Klonoff-Cohen et al., 1989). After controlling for gravidity, marital status, smoking, and alcohol consumption during pregnancy, family history of preeclampsia, working status during pregnancy, and history of hypertension in the subjects' mothers, they found that single women who used barrier contraception had a 2-fold higher risk of developing preeclampsia. A dose–response effect was also observed, suggesting that the risk of preeclampsia increased as exposure to sperm and seminal fluid decreased.

The effect of semen familiarity may not be confined to the reproductive tract. In a recent study, it was discovered that women who engaged in oral sex with their partners prior to pregnancy displayed significantly lower rates of preeclampsia (Koelman *et al.*, 2000). Among the preeclamptic women, 44% reported having oral sex with their partner, compared to 88% of those without preeclampsia. Furthermore, only 17% of the preeclamptic women, compared to 48% of the nonpreeclamptic women, reported ingesting their partner's semen.

The length of cohabitation prior to conception has been identified as another risk factor for preeclampsia. Women with an extended period of cohabitation with the child's father prior to becoming pregnant experienced a lower incidence of preeclampsia (Marti & Hermann, 1977). One account of this effect is that, other things being equal, semen familiarity would be proportional to the length of cohabitation.

A change in paternity, from one pregnancy to the next, also increases the risk of developing preeclampsia. Trupin, Simon, and Eskenazi (1996) examined a sample of 10 868 women, including 5068 nulliparous and 5800 multiparous females. Socioeconomic characteristics, reproductive and medical history, health behavior, and marital history information was obtained from a personal interview. Consistent with data from other studies implicating the importance of semen familiarity, new-partner multiparous women had a 29% higher risk of developing preeclampsia than their multiparous counterparts for whom paternity did not change.

Additional evidence also implicates changing paternity as a risk factor for preeclampsia. In a study of Caribbean women (Robillard *et al.*, 1993), 61.7% of those with preeclampsia had new-partner pregnancies, compared to only 10% among women with chronic hypertension, and 16.6% in controls. Furthermore, when they examined three or four consecutive pregnancies, rates of preeclampsia appeared to summate with each additional change in paternity. Other researchers have found similar results. In Nigeria, Ikedife (1980) documented over a period of 10 years that nearly three out of every four (74%) preeclamptic multiparous women had experienced changes in paternity.

Familiar semen as an index of paternal investment

Under conditions that prevailed hundreds of thousands of years ago, without a committed male partner to provide protection and care, the costs of reproduction for females would have increased dramatically. The existence of a maternal mechanism that could distinguish between familiar and unfamiliar semen, as a means of distinguishing between committed and less committed males, would have had considerable adaptive significance. Thus, the application of evolutionary meta-theory may provide insight into the etiology of preeclampsia. It may be useful to think about preeclampsia not simply as a medical anomaly, but as an adaptation that may have evolved to terminate pregnancies where future paternal investment was questionable or unlikely.

Sexual intercourse has the potential to be very costly for females. In addition to becoming pregnant, the burden of the infant's primary source of food and care is borne principally by mammalian females. Because pregnancy precludes further impregnation and lactation suppresses ovulation (Anderson, 1983), prior to bottle feeding when a woman became pregnant the result was a 2–4 year period in which she foreclosed on other reproductive opportunities. These costs would be exaggerated in situations where there was little or no chance of receiving care, protection, and investment from the father.

Unlike many other species, humans produce very few offspring and invest heavily in each one (Bjorklund & Shackelford, 1999; Geary, 2000). Human infants are more reliant on their parents for protection and provisioning over longer periods than any other species (Bjorklund, 1997). During human evolutionary history, in order for offspring to survive, investment of considerable time and resources was needed from both parents. Because the burden of child rearing falls more heavily on females, evolutionary pressures may have operated on women to distinguish between potential mates according to whether they were likely to enter into long-term, committed provisioning and protective arrangements with the female and her offspring (Buss, 1994).

Conception as a result of nonconsensual intercourse, or rape, represents an extremely costly pregnancy. It is highly unlikely the child's father will care for the mother and her infant. Moreover, women who conceive a child as a result of rape often face abandonment from their current partner and a reduced chance of attracting committed mates in the future (Shields & Shields, 1983; Thornhill & Thornhill, 1983, 1990). Forced copulation also precludes exercising mate choice; females have an important vested interest in the other 50% of the genes being carried by each of their offspring. As a result, it would be in a female's best interests to minimize the chances of conception as a consequence of rape. Indeed, there is recent evidence that human females may have an ensemble of evolved strategies that function to reduce the likelihood of being raped during the ovulatory phase of the menstrual cycle when they would be more likely to conceive.

Chavanne and Gallup (1998) found that instances of sexual risk-taking behavior were inversely proportional to the chance of becoming pregnant. Female college students who were in the ovulatory phase of their menstrual cycle were less likely to go out alone, walk in dimly lit areas, or go drinking in bars. In a follow-up study, handgrip strength (as an indirect measure of a woman's ability to resist sexual assault) also varied as a function of the menstrual cycle. After reading a passage containing sexually coercive overtones, females in the ovulatory phase (but not other phases) showed significant increases in handgrip strength (Petralia & Gallup, 2002). Taken together, these studies suggest that women may have been selected during human evolutionary history to behave in ways that reduce the likelihood of conception as consequence of rape, a situation in which the chances of future paternal provisioning are slim to none.

However, should conception occur following rape, a mechanism that could distinguish between pregnancies as a function of the likelihood of future investment from the father would have considerable adaptive value. In our view preeclampsia may be one of several candidate mechanisms that evolved to reduce the likelihood that such pregnancies would go full term.

Categories of unfamiliar semen

One simple algorithm that may have evolved to solve the adaptive problem posed by both the mother's and infant's need for a caring, committed adult male would be a mechanism that could distinguish between familiar and unfamiliar semen. Frequent insemination of a female by the same male over an extended period of time would be a relatively good biochemical index of the existence of a committed pair bond and, therefore, semen familiarity would predict the likelihood of long-term provisioning, protection, and care of the mother and the child during pregnancy and following parturition.

There are at least three categories of unfamiliar semen that define situations in which it might be in the female's reproductive interests to terminate a pregnancy. Forced copulation or rape is usually represented by a single instance of exposure to unfamiliar semen. Should conception occur as a result of being raped the likelihood of assistance and commitment by the father is negligible. In terms of a single exposure to unfamiliar semen, artificial insemination represents the closest approximation to rape. Semen unfamiliarity also characterizes dishonest mating strategies (Shields & Shields, 1983). For example, it is not uncommon for males to feign good intentions and commitment to females as a means of attempting to gain sexual access (Tooke & Camire, 1991). Should conception occur under such circumstances, the likelihood of paternal investment is low. Finally, semen unfamiliarity also characterizes the early phases of honest courtship in which the status of the relationship has yet to be determined. Should pregnancy occur prior to the development of a strong, committed bond between the male and female, the likelihood of abandonment by the father may still be relatively high.

In each of these situations, semen familiarity can be used to assay the like-lihood of future paternal investment. A mechanism (i.e. a preeclampsia trigger) that could distinguish between familiar and unfamiliar semen as a means of differentiating between committed and uncommitted males could have had considerable adaptive significance.

Recurrent insemination by the same male not only serves as an indicator of the likelihood of future paternal investment. After conception occurs continued insemination by the father may have other adaptive consequences. The cyto-kines interleukin 6 and 8 and tumor necrosis factor found in semen (Maegawa et al., 2002) are involved with mechanisms that regulate placentation, implanta-tion, and fetal maturation. Likewise, the high levels of prostaglandins in semen may diminish the immune response of the mother to the fetus. The presence of prostaglandins, placental hormones 5, 12, and 14, human chorionic gonadotro-pin, luteinizing hormone, and follicle-stimulating hormone in human semen (see Chapter 8 in this volume) are all components that have been implicated in preeclampsia (Ney, 1986; Pridjian and Puschett, 2002; Seppälä et al., 1985). Therefore, the absence of recurrent insemination by the father during the initial stages of pregnancy may lead to a series of events that eventuate in the failure to achieve implantation or spontaneous abortion. Consistent with the fact that humans are the only species to undergo a second phase of implantation, there may be a critical period of prenatal development in which the presence of the father's semen facilitates the second phase of implantation.

There are other etiological factors involved in preeclampsia and ours is not the only evolutionary account. Takiuiti, Kahale, and Zugaib (2003) have advanced the hypothesis that preeclampsia is a maladaptation to stress. They argue that because pregnant females are more vulnerable to predation as a consequence of being slower, heavier, sleepier, and less agile, higher blood pressure during pregnancy and a heightened alarm response may minimize the risk of predation. However, this hypothesis fails to explain why human females are the only mammals that are prone to preeclampsia, and it does not account not for the evidence implicating semen unfamiliarity as a risk factor for preeclampsia.

Preeclampsia can be fatal not only for fetuses, but occasionally for mothers as well (Duley, 1992). In our view, however, the costs associated with preeclampsia for the mother would typically be outweighed by the benefits that derive from terminating a costly pregnancy before the child has been born.

Other pregnancy complications

We suspect that there are two broad categories of reproductive termi-nation mechanism. The first operates primarily during the prenatal period. We have argued that preeclampsia is a biological adaptation that evolved to terminate investment in a pregnancy in which the likelihood of future investment from the father is low or when overall infant viability is low. Spontaneous abortion may be another example of evolved pregnancy-termination mechanisms. It is estimated that about 50% of all fertilized eggs are spontaneously aborted before women even know they are pregnant (Ellish et al., 1996; Garcia-Enguidanos et al., 2002). Even among known pregnancies, the spontaneous abortion rate is about 12–15% (Garcia-Enguidanos et al., 2002; Zinaman et al., 1996).

Spontaneous abortion usually results from fetal genetic abnormalities, but can also result from infection, and maternal immune responses and health problems. A study examining the risk factors for spontaneous abortion in Italy from 1978 to 1995 found that unmarried women were at an increased risk of having spontaneous abortions compared to their married counterparts (Osborne, Cattaruzza, & Spinelli, 2000). Similarly, another study focusing on late spontaneous abortions (those that occur in weeks 14–21 of pregnancy) found that women who were living alone were at greater risk of late abortions than those who were married (Ancel et al., 2000). Thus, in some instances, spontaneous abortion may also be an adaptation that terminates investment in offspring under circumstances where re-insemination by the father stops (as an indicator of abandonment) and, as a consequence, the mother faces a lack of paternal support.

The second category of termination mechanisms includes those that operate during the postnatal period. Daly and Wilson (1984, 1988) suggest that rather than a feature of abnormal psychological processes, postpartum depression may be a functional component of human parental decision-making. A review of the literature provided by Hagan (1999) supports this view. Hagan proposed the defection hypothesis, which posits that postpartum depression functions to inform mothers that they have suffered or are suffering a fitness cost. The psychological symptoms associated with depression motivate them to reduce or to terminate investment in the infant. This prediction is supported by a large body of evidence showing that postpartum depression can be triggered by a lack of social support, particularly from the father (see Hagan, [1999] for review). Consistent with this hypothesis is evidence that receiving more social support can lead to the remission of depressive symptoms (O'Hara, 1985). Thus, infanticide may be another postpartum psychological mechanism that evolved to terminate investment by the mother (Daly & Wilson, 1984, 1988). As an extension of our hypothesis, it would be interesting to see if the probability of postpartum depression or infanticide is related to the frequency with which the mother is exposed to the father's semen during both the pre- and postnatal period.

Conclusion

Because the costs of reproduction for females are so high and human infants require extended care and provisioning, increasingly during human evolution successful reproduction (i.e. producing children that live long enough to produce children of their own) came to depend on parental investment by both the mother and the father. Females without committed, caring male partners would have been at an enormous disadvantage when it came to child-bearing and child-rearing. We theorize that because of the growing importance of paternal investment, mechanisms may have evolved to terminate pregnancies under conditions in which support and provisioning by the child's father were doubtful. One reliable means of indexing paternal commitment would have been frequent and recurrent insemination of the female by the child's father. Subtle differences between males in semen chemistry could have been the basis for the evolution of an ensemble of pregnancy termination mechanisms triggered by impregnation as a byproduct of exposure to unfamiliar semen.

References

Amato, P. R. and Rivera, F. (1999). Paternal involvement and children's behavior problems. *Journal of Marriage and Family*, **61**, 372–84.

Ancel, P. Y., Saurel-Cubizolles, M. J., DiRenzo, G. C., Papiernik, E., and Breart, G. (2000). Risk factors for 14–21-week abortions: a case control study in Europe. The Europop group. *Human Reproduction*, **15**, 2426–32.

Anderson, P. (1983). The reproductive role of the human breast. *Current Anthropology*, **24**, 25–32.

Astin, M., Scott, J. R., and Worley, R. J. (1981). Pre-eclampsia/eclampsia: a fatal father factor. *Lancet*, **2**, 533.

Bellinge, B. S. Copeland, C. M., Thomas, T. D., *et al.* (1986). The influence of patient insemination on the implantation rate in an in vitro fertilization and embryo transfer program. *Fertility and Sterility*, **46**, 252–6.

Bjorklund, D. F. (1997). The role of immaturity in human development. *Psychological Bulletin*, **122**, 153–69.

Bjorklund, D. F. and Shackleford, T. K. (1999). Differences in Parental Investment contribute to important differences between men and women. *Current Directions in Psychological Science*, **8**, 86–9.

Buss, D. M. (1994). What do people desire in a mate? The evolution of human sexual strategies. *Journal of NIH Research*, **6**, 37–40.

Chavanne, T. J. and Gallup, G. G., Jr. (1998). Variation in risk taking strategies as a function of the menstrual cycle. *Evolution and Human Behavior*, **19**, 27–32.

Chng, P. K. (1982). Occurrence of pre-eclampsia in pregnancies to three husbands. Case report. *British Journal of Obstetrics and Gynaecology*, **10**, 862–3.

Cunnane, S. C., Harbige, L. S., and Crawford, M. A. (1993). The importance of energy and nutrient supply in human brain evolution. *Nutrition and Health*, **9**, 219–35.

Daly, M. and Wilson, M. I. (1984). A sociobiological analysis of human infanticide. In G. Hausfater and S. B. Hrdy, eds., *Infanticide: Comparative and Evolutionary Perspectives*. New York: Aldine, pp. 487–502.

Daly, M. and Wilson, M. (1988). *Homicide*. New York: Aldine de Gruyter.

Dekker, G. A., Robillard, P. Y., and Hulsey, T. C. (1998). Immune maladaptation in the etiology of preeclampsia: a review of corroborative epidemiologic studies. *Obstetrical and Gynecological Survey*, **53**, 377–82.

Dornbusch, S., Carlsmith, J. M, Bushwall, S. J., *et al.* (1985). Single parents, extended households, and the control of adolescents. *Child Development*, **56**, 326–41.

Duley, L. (1992). Maternal mortality associated with hypertensive disorders of pregnancy in Africa, Asia, Latin America and the Carribean. *British Journal of Obstetrics and Gynaecology*, **99**, 547–53.

Ellish, N. J., Saboda, J. O., O'Connor, P. C., *et al.* (1996). A prospective study of early pregnancy loss. *Human Reproduction*, **11**, 406–12.

Fagan, J. and Iglesias, A. (1999). Father involvement program effects on fathers, father figures, and their head start children: a quasi-experimental study. *Early Childhood Research Quarterly*, **14**, 243–69.

Feeney, J. G. and Scott, J. S. (1980). Pre-eclampsia and changed paternity. *European Journal of Obstetrics and Gynecology*, **1**, 35–8.

Flouri, E. and Buchanan, A. (2002). Father involvement in childhood and trouble with the police in adolescence: findings from the 1958 British birth cohort. *Journal of Interpersonal Violence*, **17**, 689–701.

Flouri, E. and Buchanan, A. (2003). The role of father involvement in children's later mental health. *Journal of Adolescence*, **26**, 63–78.

Friedman, S. A. (1993). Hypertensive disorders of pregnancy. In J. S. Brown, and W. R. Andrombleholme, eds., *Handbook of Gynecology & Obstetrics, 1st edn.* Norwalk, CT: Appleton and Lange, pp. 405–16.

Garcia-Enguidanos, A., Calle, M. E., Valero, J., Luna, S., and Dominguez-Rojas, V. (2002). Risk factors in miscarriage: a review. *European Journal of Obstetrics and Gynecology and Reproductive Biology*, **102**, 111–19.

Gaudino, J. A., Jr., Jenkins, B., and Rouchat, R. W. (1999). No father's names: a risk factor for infant mortality in the state of Georgia. *Social Science and Medicine*, **48**, 253–65.

Geary, D. C. (2000). Evolution and proximate expression of human paternal investment. *Psychological Bulletin*, **126**, 55–77.

Hack, M., Klein, N. K., and Taylor, H. G. (1995). Long-term developmental outcomes of low birth weight infants. *Future of Children*, **5**, 176–96.

Hagan, E. H. (1999). The functions of post-partum depression. *Evolution and Human Behavior*, **20**, 325–59.

Haig, D. (1993). Genetic conflicts in human pregnancy. *Quarterly Review of Biology*, **68**, 495–532.

Hoy, J., Venn, A., Halliday, J., Kovacs, G., and Waalwyk, K. (1999). Perinatal and obstetric outcomes of donor insemination using cryopreserved semen in Victoria, Australia. *Human Reproduction*, **14**, 1760–4.

Hwang, C. P. and Lamb, M. E. (1997). Father involvement in Sweden. A longitudinal study of its stability and correlates. *International Journal of Behavioral Development*, **21**, 621–32.

Ikedife, D. (1980). Eclampsia in multipara. *British Medical Journal*, **6219**, 985–6.

Jones, M. E. and Bond, M. L. (1999). Predictors of Birth outcome among Hispanic immigrant women. *Journal of Nursing Care Quality*, **14**, 56–62.

Klonoff-Cohen, H. S., Savitz, D. A., Cefalo, R. C., and McCann, M. F. (1989). An epidemiologic study of contraception and preeclampsia. *Journal of the American Medical Association*, **262**, 3143–7.

Koelman, C. A., Coumans, A. B., Nijman, H. W., *et al.* (2000). Correlation between oral sex and low incidence of preeclampsia: a role for soluble HLA in seminal fluid? *Journal of Reproductive Immunology*, **46**, 155–66.

Mac Gillivray, I. (1983). Factors predisposing to the development of preeclampsia. In I. Mac Gillivray, ed., *Pre-eclampsia. The Hypertensive Disease of Pregnancy*. London: W.B. Saunders, pp. 23–55.

Maegawa, M., Kamada, M., Irahara, M., *et al.* (2002). A repertoire of cytokines in human seminal plasma. *Journal of Reproductive Immunology*, **54**, 33–42.

Marconi, G., Auge, L., Oses, R., *et al.* (1989). Does sexual intercourse improve pregnancy rates in gamete intrafallopian transfer? *Journal of Fertility and Sterility*, **51**, 357–9.

Marti, J. J. and Herrmann, U. (1977). Immunogestosis: a new etiologic concept of "essential" EPH gestosis, with special consideration of the primigravid patient; preliminary report of a clinical study. *American Journal of Obstetrics and Gynecology*, **128**, 489–93.

Martin, R. D. (1996). Scaling of the mammalian brain: the maternal energy hypothesis. *News in Physiological Sciences*, **11**, 149–56.

Mulkey, L. M., Crain, R. L., and Harrington, A. J. C. (1992). One-parent households and achievement: Economic and behavioral explanations of a small effect. *Sociology & Education*, **65**, 48–65.

Mutale, T., Creed, F., Maresh, M., and Hunt, L. (1991). Life events and low birthweight – analysis by infants preterm and small for gestational age. *British Journal of Obstetrics and Gynaecology*, **98**, 166–72.

Need, J. A., Bell, B., Meffin, E., and Jones, W. R. (1983). Pre-eclampsia in pregnancies from donor inseminations. *Journal of Reproductive Immunology*, **5**, 329–38.

Ney, P. G. (1986). The intravaginal absorption of male generated hormones and their possible effect on female behaviour. *Medical Hypotheses*, **20**, 221–31.

Oakley, A. (1985). Social support and the outcome in pregnancy: the soft way to increase birth weight? *Social Science and Medicine*, **21**, 1259–68.

O'Hara, M. W. (1985). Depression and marital adjustment during pregnancy and after delivery. *American Journal of Family Therapy*, **13**, 49–55.

Osborne, J., Cattaruzza M. S., and Spinelli, A. (2000). Risk of spontaneous abortion in Italy, 1978–1995, and the effect of maternal age, gravidity, marital status, and education. *American Journal of Epidemiology*, **151**, 98–105.

Padilla, Y. C. and Reichamn, N. E. (2001). Low birthweight: do unwed fathers help? *Children and Youth Services Review*, **23**, 427–52.

Petralia, S. M. and Gallup, G. G., Jr. (2002). Effects of a sexual assault scenario on handgrip strength across the menstrual cycle. *Evolution and Human Behavior*, **23**, 3–10.

Pipkin, F. and Rubin, P. C. (1994). Pre-eclampsia – the "disease of theories". *British Medical Bulletin*, **50**, 381–96.

Pridjian, G. and Puschett, J. B. (2002). Preeclampsia. Part 2: experimental and genetic Considerations. *Obstetrical and Gynecological Survey*, **57**, 619–40.

Ramsey, C. N., Abell, T. D., and Baker, L. C. (1986). The relationship between family functioning, life events, family structure, and the outcome of pregnancy. *Journal of Family Practice*, **22**, 521–7.

Robillard, P. Y., Hulsey, T. C., and Alexander, G. R. (1980). Paternity patterns and risk of preeclampsia in the last pregnancy in multiparae. *Journal of Reproductive Immunology*, **24**, 1–12.

Robillard, P. Y., Hulsey, T. C., Alexander, G. R., *et al.* (1993). Paternity patterns and risk of preeclampsia in the last pregnancy in multiparae. *Journal of Reproductive Immunology*, **24**, 1–12.

Robillard, P. Y., Dekker, G. A., and Hulsey, T. C. (2002). Evolutionary adaptations to pre-eclampsia/eclampsia in humans: low fecundability rate, loss of oestrus, prohibitions of incest and systematic polyandry. *American Journal of Reproductive Immunology*, **47**, 104–11.

Robillard, P. Y., Hulsey, T. C., Dekker, G. A., and Chaouat, G. (2003). Preeclampsia and human reproduction: an essay of a long term reflection. *Journal of Reproductive Immunology*, **59**, 93–100.

Scott, J. R. and Beer, A. A. (1976). Immunologic aspects of pre-clampsia. *American Journal of Obstetrics and Gynecology*, **125**, 418–27.

Seppälä, M., Koskimies, A. I., Tenhunen, A., *et al.* (1985) Pregnancy proteins in seminal plasma, seminal vesicles, preovulatory follicular fluid, and ovary. *Annals of the New York Academy of the Sciences*, **442**, 212–26.

Shields, W. M. and Shields, L. M. (1983). Forcible rape: an evolutionary perspective. *Ethology and Sociobiology*, **4**, 115–36.

Smith, G. N., Walker, M., Tessier, J. L., and Millar, K. G. (1997). Increased incidence of preeclampsia in women conceiving by intrauterine insemination with donor versus sperm for treatment of primary infertility. *American Journal of Obstetrics and Gynecology*, **177**, 455–8.

Starfield, B., Shapiro, J., Weiss, K., *et al.* (1992). Race, family income, and low birthweight. *American Journal of Epidemiology*, **134**, 1167–96.

Takakuwa, K., Honda, K., Ishii, K., *et al.* (1999). Studies on the HLA-DRB1 genotypes in Japanese women with severe pre-eclampsia positive and negative for anticardiolipin antibody using a polymerase chain reaction-restriction fragment length polymorphism method. *Human Reproduction*, **12**, 2980–6.

Takiuiti, N. H., Kahale, S., and Zugaib, M. (2003). Stress-related preeclampsia: an evolutionary maladaptation in exaggerated stress during pregnancy? *Medical Hypotheses*, **60**, 328–31.

Thaler, C. J. (1989). Immunological role for seminal plasma in insemination and pregnancy. *American Journal of Reproductive Immunology*, **21**, 147–50.

Thornhill, R. and Thornhill, N. W. (1983). Human rape: an evolutionary analysis. *Ethology and Sociobiology*, **4**, 137–73.

Thornhill, N. W. and Thornhill, R. (1990). Evolutionary analysis of psychological pain of rape victims I: the effects of victim's age and marital status. *Ethology and Sociobiology*, **11**, 155–76.

Tooke, W. and Camire, L. (1991) Patterns of deception in intersexual and intrasexual mating strategies. *Ethology and Sociobiology*, **12**, 345–64.

Turner, J. R., Grindstaff, C. F., and Phillips, N. (1990). Social support and outcome in teenage pregnancy. *Journal of health and Social Behavior*, **31**, 43–58.

Trupin, L. S., Simon, L. P., and Eskenazi, B. (1996). Change in paternity: a risk factor for preeclampsia in multiparas. *Epidemiology*, **7**, 240–4.

Yongman, M. W., Kindlon, D., and Earls, F. (1995). Father involvement and cognitive/behavioral outcomes of preterm infants. *Journal of the American Academy of Child and Adolescent Psychiatry*, **34**, 58–66.

Zinaman, M. J., Clegg, D. E., Brown, C. C., O'Connmor, J., and Selevan, S. G. (1996). Estimates of human fertility and pregnancy loss. *Fertility and Sterility*, **65**, 503–9.

PART IV   ASSESSING PATERNITY: THE ROLE
          OF PATERNAL RESEMBLANCE

11

The effect of perceived resemblance and the social mirror on kin selection

REBECCA L. BURCH, AND DANIEL HIPP
State University of New York at Oswego
AND
STEVEN M. PLATEK
Drexel University

Paternal resemblance

Due to the asymmetric risk of cuckoldry, assurance of paternity is much less significant than maternal assurance. Females, possessing 100% assurance of their link to their offspring, necessarily have evolved strategies to secure the less sure males as supportive and protective fathers. Hofferth and Anderson (2003) compared all types of family and the paternal investment inherent in each. The least-investing type of father was the stepfather. Daly and Wilson (1982) recorded spontaneous remarks in maternity wards regarding the appearance of newborn children. Mothers and their friends and relatives were more likely to comment on how children resembled their fathers than they were to say the child resembled the mother or any other family member. When fathers displayed any doubt, the mothers were quick to reassure them of the child's resemblance. Regalski and Gaulin (1993) have replicated these findings using Mexican families. These researchers concluded that women and their families attempt to reassure the male of this paternity, thus increasing the likelihood that he will invest in the child.

It is obvious that convincing a male of paternity and securing his investment would almost always be in the best evolutionary interests of females. However, this is hardly in the best interests of males. If ascriptions of resemblance were completely persuasive throughout evolutionary history, males would have been deceived numerous times into investing in a child that was not genetically related to them. Indeed, the incidence of cuckoldry ranges from 5 to 30% (see Baker & Bellis [1995] for review). Those males who, quite literally, blindly trusted their female partners and invested in unrelated children would have had their genes eradicated from the gene pool. Those males who remained wary and used their own perceptions of resemblance and invested accordingly would have stood a better chance of passing on their genetic material.

McLain *et al.* (2000) investigated this idea by objectively testing new mothers' ascriptions of paternal resemblance. Maternal ascriptions of resemblance could not be verified by the objective, unrelated raters. In no case were the infants' pictures matched to the fathers' photos more often; the mothers' opinions held no validity. These findings would, in turn, explain the reluctance in the males to prematurely agree with their partner's assertions of paternity (Daly & Wilson, 1982; Regalski & Gaulin, 1993). In some cases these males would not agree after several maternal attempts to persuade them. One of the most interesting cases depicted by Daly and Wilson (1982) was of a man who made it very clear to the hospital staff that if the child did not resemble him (regardless of others' opinions), he would abandon it and the mother at that moment.

According to Daly and Wilson (1998) there are two ways in which a male can increase the probability that the children he is caring for are carrying his genetic material: he can monitor and/or sequester the female partner during the period that she is fertile to minimize being cuckolded, or he can attempt to assess paternity based on the degree to which the children resemble him. There are two forms that a paternal-resemblance mechanism might take: (1) a social mirror mechanism (i.e. the degree to which a male is told a child resembles him) and/or (2) the degree to which the child actually resembles him.

In a population of males convicted of domestic violence, Burch and Gallup (2000) describe and illustrate the social mirror concept thoroughly. Fifty-five men in a treatment program for domestic violence offenders in upstate New York were given a questionnaire that included questions on the types of abuse they had committed, the relationships, degree of relatedness, and perceived resemblance they shared with their children. After the questionnaire was collected, the severity of abuse and degree of injury incurred as a result of the abuse was evaluated by independent raters. Burch and Gallup (2000) found that degree of perceived paternal resemblance had a significant positive correlation with a number of variables. Relationships between the males and biologically related children were rated significantly better than relationships with unrelated children. The abusive males also had worse relationships, more abuse, and increased severity of abuse with their mates as perceived resemblance to the offspring scores decreased. It was in this study that the social mirror effect was mentioned. "This social affirmation of paternal resemblance demonstrates what could be called a 'social mirror' effect. Prior to the construction of mirrors that would permit a more explicit means of evaluating phenotypic resemblance, men could have relied on this 'social mirror,' as represented by the opinions of their kin and community members (Daly & Wilson, 1982; Regalski & Gaulin, 1993 [as cited by Burch & Gallup, 2000]) to assess paternity and modify their paternal investment accordingly." (Burch & Gallup, 2000, p. 434).

Although these data indicate that both paternal resemblance and social affirmation of resemblance play a role in the treatment of offspring, they are also quite limited. For example, the data are correlational in nature; it is possible that the relationship with the child influences ascription, and not vice versa. Secondly, other situational factors may influence both the father's ratings of child treatment and his ascription of resemblance. This study also did not compare the respective contributions that personal ascription of resemblance and social ascription of resemblance of the child made to the treatment of the child or spouse.

Another question that arises from this analysis is, how confident can a man be in his gauging of resemblance in his putative offspring? Surely other factors, such as medical conditions (deformities, jaundice, changes in skin coloration at birth) or interpersonal issues (suspected infidelity of the partner, behavioral problems in the child) could skew a man's opinion.

Platek et al. (2002) studied the dimorphic reactions males and females had when presented with digitally morphed images of children. The children's pictures were morphed with images of the subjects. To measure reactions, the researchers presented the subjects with hypothetical parental investment situations, asking which of a group of 10 children, including one picture morphed with their own image, the subject would be most likely to choose. Males chose their own morphed picture more often than females in almost every positive category and less often in almost every negative category. The greatest difference in male and female responses occurred in the question, "Which one of these children would you be most likely to adopt?" According to Platek et al. (2002), "none of the males picked children's faces that resembled their own as the child they would be the most likely to punish." The researchers also went on to predict that maternal attempts at convincing the father of paternity would correlate positively with the mother's infidelity.

Recognizing the correlative significance of paternal resemblance with paternal investment, Platek et al. (2003) sought to determine whether males were more capable of determining resemblance or whether it was simply the evolutionarily significant situation of paternal investment that triggered a more analytical look at resemblance. The data were conclusive. Males and females were equally able to determine resemblance. This indicated that it was the nature of the hypothetical investment questions that caused the sexually dimorphic responses. Platek and colleagues also found that as the percentage of characteristics shared with the subject increased from 12.5 to 50.0%, the amount of investment on the part of the males increased accordingly. This demonstrates that this differential investment is not a simple on/off response, but fluctuates with the magnitude of resemblance. This has obvious implications for investment in different family members of varying relatedness.

Platek *et al.* (2004; and Chapter 12 in this volume) used functional magnetic resonance imaging to determine the neurobiological correlates associated with processing facial resemblance that were predicted to be a proxy for detection of paternity. When presented with pictures of self-child morphs, males showed activation in the left frontal part of the brain (left superior, middle, and medial frontal gyri), while females showed right-brain activation (right superior and medial frontal gyri). There was no difference in activation when presented with non-self-morphs. This suggests that a specific mechanism that provides for an increased ability to discern paternity exists in males, and is located specifically in the left superior, middle, and medial frontal gyri. This would have to be present in an ancestral environment for any differing ability to occur, as male–infant and female–infant resemblance is normally distributed, with only 20% of infants resembling either parent in any significant measure (Bressan & Grassi, 2004).

The social mirror effect was again displayed with even more subjects, this time in the UK by Apicella and Marlowe (2004). The researchers sought to further investigate the role of perceived maternal fidelity and paternal resemblance as they relate to paternal investment. Previous studies (Buss, 1989; Buss & Schmitt, 1993) had shown that males valued chastity much more significantly than females because of its significance in cuckoldry prevention. Predictably, it was also found that ratings of perceived maternal fidelity decreased significantly when the couple had broken up at any time. This significant body of evidence points to the idea that the human brain possessed the modules to attempt to recognize its closest genetic relatives even before it provided itself with the luxury of an actual mirror.

The social mirror

The mere fact that a man can ascribe resemblance must be premised on the fact that he has seen his own face. Humans, social groups, and cuckoldry existed prior to the advent of mirrors. During evolutionary history, due to the lack of ostensible mirrors with which to view one's own appearance, males might have been selected to rely heavily on the ascriptions of resemblance by those of their social group.

Although it appears adaptive for both the females to ascribe resemblance and for the males to be suspicious, it remains to be seen just how much this social mirror truly affects male's ascriptions of resemblance and, ultimately, their investment in putative offspring.

The interaction of actual resemblance and social mirror-mediated resemblance ascription has yet to be tested objectively. To do this, S. M. Platek and colleagues (unpublished data) recruited 20 (10 male, 10 female) undergraduates. Each subject's picture was morphed with the image of a female and a male 2-year-old child so that the image was 50% of the subject's face and 50% of the child's face. Much like prior morphing studies, each trial consisted of the subject being shown five faces and presented with a question in the middle of the array (e.g. "Which of these children would you spend the most time with?", "Which one of these children would you be most likely to adopt?", etc.). In some conditions, the subjects were told that the computer compared their picture to a database of pictures and generated feedback as to whether any of the children's faces shared facial characteristics with them. In the first social mirror condition the subjects were given information by the computer that the face they were morphed with did share characteristics with them. In the second social mirror condition they were told that the face they were morphed with did not. In the second and third social mirror conditions the subjects were told that the face of another person morphed with the child either did (condition 3) or did not (condition 4) share facial characteristics with them. Each face array was shown with each of 10 questions for a total of 50 stimulus trials. Whether the subject saw the faces morphed with the girl or boy child's face in the social mirror conditions was randomized across subjects. The face presentation was randomized across trials. Latency to respond was recorded for all questions. At the end of the experiment subjects were asked how they made their choices (open-ended) and how difficult it was to make a choice on a scale of 1–5 (1 being the easiest and 5 being the most difficult).

There were no main effects for the sex of the toddler face. Similar to our previous findings (Platek *et al.*, 2002), when given no social mirror feedback males were more likely to choose faces they had been morphed with than were females for positive-type questions. Males were significantly more likely to choose their own face morph in response to "Which one of these children would you be most likely to adopt?" and "Which one of these children do you think that you would spend most time with?". Unlike males, there were no questions where females were more likely to choose their own face/toddler morph. In fact, females chose no face, or face position coordinate, more often than chance. In the second array of faces where the subject's face was not morphed with any of the child faces, there were no sex differences in the likelihood to select any particular face.

Pairwise *post hoc* comparisons revealed that when a self-morph face was said to *not* share characteristics with the subject and a non-self-morph was said to share characteristics with the subject, males selected their own face morph less often. Positive social mirror ascriptions to a self-morph did not raise reactions above those when there was no feedback; i.e., in both conditions males selected their own face morph 80% of the time. In other words, when males were given feedback by the computer that conflicted with their perceptions of resemblance, this did affect their investment choices, but investment did not change when the males' suspicions are affirmed. However, there was also a social mirror condition by gender interaction. Pairwise *post hoc* comparisons revealed that males reacted to their own face morphs more positively than females in conditions 1 (self-morph said to resemble) and 4 (non-self-morph said not to resemble), so males are still investing more than females even when the computer is affirming suspicions; females are still choosing their morphs less, even with affirmation from the computer. This implies, as one would suspect, that resemblance, and also social mirror, have little effect on female investment choices.

Specifically for the social mirror condition of "non-self-morph said to resemble the subject," we examined the amount of times the non-self-morph was selected by males. For the positive-type questions this face was not selected a great amount of the time (mean = 0.7, S.E.M. ± 0.3958). Likewise, we examined the number of times other faces were selected when the self-morph was said *not* to resemble the subject; there was no trend toward males selecting one other face more often than the others. This implies that while males will take social mirror into consideration, they will remain wary of investing in that child. This illustrates in an experimental setting the same reluctance in males found in Daly and Wilson (1982) and Regalski and Gaulin (1993). In summary, males select a face primarily by using actual resemblance, but are affected by the social ascriptions of resemblance. The only social mirror conditions that seemed to affect reactions toward the faces were the conditions in which the subject was told that a self-morph did not resemble him and that a non-self-morph did resemble him. This seems to represent a situation in which sensory information about resemblance is being weighed in terms of males' reactions toward the faces.

The relationship between social mirror-mediated resemblance and actual resemblance is complex. Actual and social mirror-mediated resemblance do not interact additively, but it is not because there is no interaction between the two processes. It seems that social mirror information only affects a male's reactions when information from the actual and the social mirror domains are incompatible. However, this information does not draw responses as one would suspect to the face that is being labeled to resemble. Therefore, tentatively we conclude that the latter prediction is probably at work; i.e. even though males were affected by verbal information about resemblance, because males were not drawn more to the labeled faces in the two incongruous social mirror conditions it seems as though males were not using that information to guide their choices.

The evolutionary basis for these findings can be found in McLain *et al.* (2000). During evolutionary history, females may have adopted a strategy in which ascribing resemblance may have resulted in increased investment from males early during human evolutionary history. Females could then utilize that information to increase their chances for successful extra-pair copulation; that is, cuckoldry. Under these conditions males would have been selected for a deception-detecting mechanism aimed to limit cuckoldry by becoming suspicious about female's ascriptions of resemblance.

The social mirror and parental treatment in college students

Although it appears adaptive for both the females to ascribe resemblance and for the males to be suspicious, it remains to be seen just how much this social mirror truly affects males' ascriptions of resemblance and, ultimately, their investment in putative offspring in real-life situations.

Using a questionnaire and a sample of 555 male and female undergraduates, we investigated the interaction of actual resemblance and social mirror-mediated resemblance ascription (R. L. Burch, unpublished data). It is important to note that this study did not investigate parents; these data are from the child's point of view. Because of this, ratings of resemblance may not be as important in parental investment, as child and parent ratings of resemblance may not be equivalent. However, social mirror ratings may be equivalent, as in most cases both child and parent are present in order for social ascriptions to be made.

First of all, males did not report that they looked significantly more like their fathers, but females reported that they looked significantly more like their mothers. However, how often others told the mother or father that the child resembled them did not differ by gender and the frequency that the parent stated that the child resembled them did not differ by gender.

Treatment of the child was categorized by overall investment, time spent with the child, time spent talking with the child, money and presents given, preferential treatment of the child, and relationship quality. Ratings of paternal and maternal resemblance did correlate to some extent with treatment of the child. However, paternal resemblance had a greater effect than maternal resemblance. For example, correlations between paternal resemblance and treatment were above $r = 0.150$ for overall investment, time spent talking, presents given, and relationship quality. Maternal resemblance failed to produce such an effect.

Interestingly, the social mirror demonstrated a much larger effect on child treatment and a much bigger gender difference. The paternal social mirror correlated with overall investment ($r = 0.210$), time spent ($r = 0.214$), time spent talking ($r = 0.251$), money given ($r = 0.185$), presents given ($r = 0.236$), preferential treatment ($r = 0.215$), and relationship quality ($r = 0.258$). For mothers, the social mirror affected investment ($r = 0.183$) and time spent talking ($r = 0.211$), but no other correlations exceeded $r = 0.150$.

When asked how their birth or adoption affected their parents' marriage, only one type of resemblance had an effect. Only the paternal social mirror correlated with quality of marriage. The more the father heard that this child resembled him, the more positively the child affected the marriage in the categories of trust ($r = 0.161$), companionship ($r = 0.173$), and overall positive effect ($r = 0.177$).

This social affirmation of paternal resemblance demonstrates what would appear to be a social mirror effect on child treatment. The social mirror answers the question of how paternal resemblance could be selected for in an ancestral setting, where there were no mirrors for the fathers to see their reflection. Without an idea of his own features, how would it be possible for a father to judge his resemblance in his ostensible children? Prior to the advent of mirrors, males could have relied on this social mirror – the opinions of their kin and community members – to determine paternity and modify their paternal investment accordingly. This raises an interesting question in light of the research by Daly and Wilson, Regalski and Gaulin, and McLain and colleagues; who does the male listen to regarding ascribed resemblance? As previous research shows, males are wary of resemblance ascription by mothers and maternal family members, and rightly so, as McLain and colleagues have shown that these ascriptions have no basis and objective raters cannot match the child's picture to the father's. The prediction would be that putative fathers would weight the ascriptions of their relatives more heavily, particularly ascriptions by closely related male relatives; brothers, for example. These relatives, by being more closely related, would not only be in a better position to rate resemblance (through familiarity with the father's face and memory of how he looked as a child), but would also be more likely to invest in this child and have a greater genetic stake in the father's investment of this child, and therefore would be similarly wary of ascribing resemblance and suggesting investment. It would be in the best interests of males to be more willing to trust brothers in resemblance ascriptions prior to the advent of mirrors, and this would be expected to still hold today, as paternal resemblance has not diminished in importance (G. G. Gallup, Jr., personal communication). If the computer-based social mirror (S. M. Platek *et al.*, unpublished data) served as an objective stranger in ascribing resemblance, the effect of paternal family members ascribing (or not ascribing) resemblance should have a much greater effect. Consistent with these findings are recent findings of Shackelford and his colleagues (2005), who showed that among divorced couples, social mirror-based resemblance perceptions significantly predicted a male's adherence to child-support obligations.

Extension of paternal resemblance: the impact on siblings

While research has been conducted on the treatment of stepchildren and the effect of parental resemblance on child treatment, few studies have examined the effect of relatedness or resemblance on sibling relationships. Jankowiak & Diderich (2000) found that the more genetically related the siblings (full versus half), the more closeness/solidarity they showed. Segal and Hershberger (1999) found that monozygotic (identical) twins cooperated more in the prisoners' dilemma paradigm than dizygotic (fraternal) twins. In an extension of the paternal resemblance literature, we posit that paternal uncertainty results in a degree of sibling uncertainty, as full siblings may only be half siblings, and half siblings may be completely unrelated. Because of this sibling uncertainty, sibling resemblance may have an effect on sibling relationships. It would also be expected that although parental resemblance has the greatest impact on child treatment, both males and females would possess some degree of sibling uncertainty and therefore there should be no gender differences in the impact of sibling resemblance on sibling rivalry.

To examine sibling uncertainty and the impact of sibling resemblance, we asked 555 undergraduate students questions about the amount of time, resources, and emotional energy that the participant invested in each sibling, resulting in a study of 963 sibling relationships (R. L. Burch, unpublished data). Although identical twins cooperated more and competed less than lesser-related siblings, the small number of dyads ($n = 3$) excluded them from analysis.

The percentage of genetic relatedness the subject shared with their siblings correlated with several variables, including ratings of resemblance between siblings, time spent with the sibling and time spent talking with the sibling, and ratings of closeness between siblings. Percentage of genetic relatedness yielded differences in several parameters of the sibling relationship. Half and full siblings got along significantly better than unrelated siblings (adopted and stepsiblings). In fact, on the positive questions regarding sibling relationships (spending time, spending time talking, and feelings of closeness) half and full siblings rated their relationships as better than unrelated siblings. On all negative questions (frequency of fighting, levels of resentment, strength of competition, severity of arguments), unrelated siblings scored higher than half siblings. Full siblings also scored higher on these items, however, possibly due to increased time spent together. Finally, both full and half siblings reported that they were favored over the adopted child or stepchild.

Resemblance had a greater effect on sibling relationships correlating with time spent with sibling ($r = 0.223$), talking with sibling ($r = 0.206$), and ratings of closeness between siblings ($r = 0.239$) and how well they got along ($r = 0.149$). Resemblance was negatively correlated with negative items such as severity of arguments or resentment.

It would be interesting to investigate not only the effect of resemblance in other parameters of the sibling relationship, but also in paradigms such as the prisoner's dilemma. It would be predicted that sibling resemblance, as measured by the siblings' opinions but also objectively rated, would have an effect on defection in social contracts, the prisoner's dilemma, and other cooperative tasks (see DeBruine [2002] for similar results). As an extension of paternal resemblance, it would also be expected that the social mirror may play a role in sibling relationships as well. For each study examining perceived resemblance, questions of social ascription should also be asked to determine which type of resemblance is having the greatest impact on sibling relationships. These questions can also be expanded to include other family members. Platek *et al.* (2003) found that the effect of paternal resemblance became nonsignificant when the faces only possessed 25% of the male's characteristics. It is possible that this effect of resemblance dwindles once it passes the 25% genetic relatedness quotient (half siblings, aunts and uncles, grandparents). However, it would be expected that extended kin identification would be necessary in an ancestral setting and may play a role even in social relationships in unrelated individuals (see below). Also, in accordance with the paternal resemblance literature, it would be interesting to examine the source of the social mirror and whether that has an effect on social mirror in siblings and other relatives.

It is important to discuss a major difference between the sibling-resemblance data and the paternal-resemblance data; that is, the lack of a gender difference. As we explained above, both males and females would be expected to be impacted by sibling uncertainty, while only males are impacted by parental uncertainty. This has interesting implications for the neurological basis for paternal resemblance versus sibling resemblance. If, using the same morphing paradigm, a gender difference is found in using resemblance to invest in children (the work of Platek and colleagues), but not in sibling investment, this implies that different brain areas or decision-making processes are being used by just changing the scenarios from "imagine these are children" to "imagine these are your siblings". It would be a large shift in decision-making based on the changing of just a few words in a hypothetical scenario! What would be the neurological basis of this shift?

As Platek *et al.* (2004) found, males showed activation in the left part of the brain (left superior, middle, and medial frontal gyri), when presented with pictures of self-morphs, while females showed right-brain activation (right superior and medial frontal gyri). Logically, one would expect if no gender difference exists in the effect of sibling resemblance, both males and females would show activation of the left superior, middle, and medial frontal gyri when asked to invest in hypothetical siblings. One would suspect that some area would be responsible for the activation of the left frontal gyri in this scenario and not in the offspring scenario in women. Of course, all of this is just speculative, but provided it is supported in the future through empirical study, implies that there are separate cognitive processes or neurological modules in determining child and sibling investment.

Each of these issues – paternal resemblance, the social mirror, sibling resemblance, etc. – have profound implications for a number of social relationships. One could argue that the problem of paternal uncertainty gave rise to the problem of sibling uncertainty, familial uncertainty, and the importance of kin recognition and the role of resemblance in kin selection. This cautiousness regarding investment in kin can easily extend into cautiousness in investing in unrelated individuals and can eventually serve as the basis of xenophobia.

Resemblance and social relationships

As resemblance plays a role in relationships in nuclear families, it would be expected that resemblance, kin identification and kin selection would be necessary even in large groups throughout the history of the human species. We hypothesize that resemblance would play a role in any number of social interactions, but particularly in situations involving altruism. Future morphing studies are intended to examine the role of resemblance in everyday social scenarios; loaning money, trust in social contracts, sharing food and resources, asking for and granting favors, etc. It would be expected that individuals would be more likely to assist those who resemble them, perhaps even rating the relationship as closer or stronger. A natural extension of this line of research would be to introduce a social mirror and determine whether this increases cooperation, trust, and altruism in similar-looking and nonsimilar-looking individuals.

This discussion of the role of resemblance in kin identification and altruism brings us to Hamilton's (1964) rule of inclusive fitness: caring for others differentially due to their genetic relatedness. Kin selection requires (1) recognition of kin and (2) the total benefit of helping a relative must exceed cost to donor's reproductive success. This is symbolized as $rb > c$ with r representing the degree of relatedness, the probability the donor and relative possess the same genes. Many have discussed this concept but relatively few have discussed the mechanism by which kin are recognized and how r would be calculated. Most of this discussion focuses on phenotype matching and olfactory processes (e.g. Lacy & Sherman, 1983). While we do not argue the role of olfactory processes in self recognition and identification of kin (see Platek, Burch, and Gallup, 2001), we posit that self-detected resemblance and the social mirror are strong factors in kin recognition in humans, and other visually based primates such as chimpanzee (F. B. M. de Waal, personal communication), with individuals capable of rating the magnitude of resemblance as seen in Platek et al. (2003). In fact, it is illustrated in detail by Platek et al. (2003) how males can determine greater or lesser resemblance in offspring and vary their investment accordingly.

A natural extension of the work of Platek and colleagues would be to morph subject faces with the faces of other adults and examine the role of resemblance and kin selection in competition and social interactions. Using the same paradigm as in the parental investment study (Platek et al., 2003), subjects would be asked to choose faces to interact with socially, share rewards with, allocate resources to, and compete or cooperate with. Social interactions could include any number of scenarios in which people must cooperate or compete with or behave altruistically toward individuals they have no prior experience with. In these scenarios, individuals must base their decisions only on the faces presented. It would be hypothesized that humans allocate more resources to and treat those that resemble them better than those that do not resemble them. Possible situations could include issues of daily social interaction such as choosing certain faces for friendship, or working or spending time together, issues of reliability or trust (loaning possession or money, allowing someone into your home, entrusting this person with secrets, doing favors for a person), sharing resources such as food, money, promotions and recognition for work, or assistance with intellectual tasks. The most dramatic situations could include scenarios regarding physically risky altruistic behavior and lifesaving. It would also be interesting to see whether subjects would seek out similar-looking faces to ask for favors, money, or generally rely on in various circumstances.

Xenophobia

Xenophobia can be defined as a fear of anything foreign, but particularly foreign people or strangers. A great deal of research has been dedicated to explaining xenophobia and racism in current cultures. Evolutionary and socio-biological explanations of ethnocentrism and xenophobia are for the most part rooted in kin selection, inclusive fitness, and altruism theories (see Reynolds, Falger, & Vine, 1987; see also discussion around race perception by Cosmides, Tooby, & Kurzban, 2003), but as stated above it may be useful to examine the role that familial uncertainty and resemblance may play in not only kin selection but also the genesis of xenophobia and outgroup negativism or hostility. Xenophobia begins with the formation of two groups, the ingroup and the outgroup. The outgroup is believed to be homogenous yet different from the ingroup and stereotypes regarding the outgroup begin to develop.

In the small bands in which humans are generally presumed to have lived during most of their evolutionary history, virtually all social interactions were among relatives. The same is probably true for contemporary hunter–gatherer societies (Van der Dennen, 1999). Throughout human evolution, these ingroups and outgroups were familial groups, more and less related, and so resemblance would be expected to be greater in the ingroup. It has been shown that kin selection plays a large role in group affiliation, as ethnic affiliation often involves some claim of common ancestry and individuals acquire a sense of kinship from the group (Van de Berghe, 1999). In this "families as groups" scenario, perceived resemblance would play a role in kin selection and the identification of the two groups. Resemblance would then play a role in levels of trust toward the outgroup, and assuming the outgroup was a collection of related and similarly looking individuals, resemblance would play a role in the stereotyping of that group.

Rydgren (2004) comments on the importance of visible markers in stereotyping and xenophobia.

> Along with sex and age, racial and ethnic characteristics are among the more immediately visible and noticeable 'social tags' that we consistently bring into our social encounters. Because racial and ethnic appearances (e.g., skin and hair color, manner of speaking, ways of dressing) are among the first pieces of information we get about people we meet, and they evoke stereotypes and prejudices, race and ethnicity have the potential for influencing our further perception of, beliefs about, and behavior toward people we meet. (Rydgren, 2004, p. 130)

Now add to these salient markers the detection of resemblance, and put these in a situation in which new individuals are encountered and predictions of their behavior must be generated. As Rydgren (2004) states, when faced with unfamiliar situations, individuals have two options: generate *a priori* predictions about a stranger's behavior, or put trust in information received from other people. Both of these options could involve the use of resemblance; *a priori* predictions based on detection of resemblance, and trust in information from others, most likely relatives, which would encompass social mirror. The extent to which either is used is determined by the advent of mirrors in the culture. This prediction can be modeled mathematically as well (see Platek, 2002).

It is important to note, as Cashdan (2001) does, that group affiliation and ingroup loyalty does not always or inevitably result in outgroup hostility. However, in situations in which there may be competition between groups for resources and survival, it would be advantageous for individuals to identify and share resources with their groups/families, notably those who resemble them, while competing against other families/groups (who do not resemble them).

Those individuals who were capable of recognizing resemblance and investing in those who resembled them invested in their kin and increased their fitness. This ability would be predicted to proliferate through a group rapidly, as an individual would be expected to pass this ability to offspring and the members of the group are, of course, related.

Ridley (1997) stated that an unexpected byproduct of the evolution of cooperative society is group prejudice. It is believed that this group prejudice and an intolerance for outgroups stems from our preference as humans to form groups (Reynolds *et al.*, 1987), but it may actually stem from our need to live with and cooperate with relatives to survive and reciprocally invest in kin. In fact, where selection pressures at the level of kin probably fostered the development of altruistic and nepotistic instincts (Trivers, 1971), the development of larger more diverse groups (possibly neighboring families) probably fostered just the opposite (Williams, 1966).

This is seen in other species, such as chimps raiding neighboring troops for females and stalking male competitors from neighboring bands (Ridley, 1997). Despite their deep roots in human history, xenophobia and the importance of resemblance detection and kin recognition are not species-specific phenomena. Chimpanzees, as our closest genetic relative, possess similar mechanisms. Parr and de Waal (1998) discovered the presence of elevated visual kin-recognition ability in chimps. Chimpanzees perceive resemblance in related but unfamiliar individuals, marking their kin recognition ability not as a mechanism relying solely on social cues but as an inherited evolutionary adaptation for visually determining a quotient of relatedness.

It is interesting to note that chimpanzees show such resemblance to kin that families can be grouped together by objective human raters (Vokey *et al.*, 2004). Parr and de Waal (1998) first found that chimpanzees can recognize kin visually, matching pictures of unfamiliar conspecifics to their offspring, specifically matching mothers to sons. The authors hypothesize that the advantages of this ability would be to prevent inbreeding, but also to assist in forming "political alliances." It is these same alliances that incur violence on neighboring groups.

Further evidence of the chimpanzee's similarly multifaceted facial recognition mechanisms is provided in a study by Parr and de Waal (1998) that investigated whether the inversion effect occurred in chimpanzee facial recognition testing. The inversion effect is defined as the impaired ability to process facial information of a familiar class of objects (e.g. humans, chimpanzees, etc.) when the observed face is inverted 180°. It was discovered that chimps, along with humans, also succumb to the inversion effect (Parr and de Waal, 1998). This is a significant finding, as it provides evidence of the relative complexity of the recognition mechanism mentioned above. Since inversion has such an effect on processing, it is clear that it is the orientation of each facial feature (to each other and to the whole of the face) that is being examined, rather than simply the features themselves. Of course it follows that this differentiation dependent on these spatial measures allows for resultant complexity in resemblance detection. However, this complexity does not occur right away. It is only after the subject crosses a critical threshold of familiarity with the class of objects being examined that the relationship of the features becomes significantly utilized in assessing resemblance. It doesn't matter if the subject is familiar with the actual object being visually observed, only that the object fits into a familiar class of objects.

So, if there is a specific visual kin-recognition mechanism as observed, and if the human's or chimpanzee's criteria used for facial analysis increases in complexity as the familiarity with the group of objects being investigated increases, it follows that the human or chimp would use their increased ability to determine resemblance as a method of determining relatedness. This is especially true of humans, since human females conceal ovulation and take more biological and behavioral steps to protect the identity of the true father of the child.

Besides having a role in determining paternity, these increased abilities may be proven to serve a function in normal social interaction. Two likely possibilities exist. Either the resemblance detection mechanism developed as a means of detecting paternity and the social interaction component evolved as a secondary function, or vice versa. We would hypothesize that paternal uncertainty gives way to sibling uncertainty, and therefore kin selection.

According to Ridley (1997), the evolution of sociability, altruism, and the instincts for coalitions goes hand in hand with hostility to outsiders. It is the advantages brought about by sociability and altruism toward kin that gives the ultimate causation to coalition building and outgroup hostility, and kin identification and selection which is the mechanism.

References

Apicella, C. L. and Marlowe, F. W. (2004). Perceived mate fidelity and paternal resemblance predict men's investment in children. *Evolution and Human Behavior*, **25**(6), 371–8.

Baker, R. and Bellis, M. (1995). *Human Sperm Competition*. London: Chapman and Hall.

Bressan, P. and Grassi, M. (2004). Parental resemblance in 1-year-olds and the Gaussian curve. *Evolution and Human Behavior*, **25**(3), 133–41.

Burch, R. L. and Gallup, G. G., Jr. (2000). Perceptions of paternal resemblance predict family violence. *Evolution and Human Behavior*, **21**(6), 429–35.

Buss, D. M. (1989). Sex differences in human mate preferences: evolutionary hypotheses tested in 37 cultures. *Behavioral and Brain Sciences*, **12**, 1–49.

Buss, D. M. and Schmitt, D. P. (1993). Sexual strategies theory: an evolutionary perspective on human mating. *Psychological Review*, **100**, 204–32.

Cashdan, E. (2001) Ethnocentrism and xenophobia: a cross cultural study. *Current Anthropology*, **42**(5), 760–4.

Cosmides, L., Tooby, J., and Kurzban, R. (2003). Perceptions of race. *Trends in Cognitive Science*, **7**, 173–9.

Daly, M. and Wilson, M. (1982). Whom are newborn babies said to resemble? *Ethology and Sociobiology*, **3**, 69–78.

Daly, M. and Wilson, M. (1998). *The Truth about Cinderella: a Darwinian View of Parental Love*. New Haven, CT: Yale University Press.

Debruine, L. M. (2002). Facial resemblance enhances trust. *Proceedings in Biological Sciences*, **269**, 1307–12.

Hamilton, W. D. (1964). The evolution of social behavior. *Journal of Theoretical Biology*, **7**, 1–52.

Hofferth, S. L. and Anderson, K. G. (2003). Are all dads equal? Biology versus marriage as a basis for paternal investment. *Journal of Marriage and Family*, **65**(1), 213–33.

Jankowiak, W. and Diderich, M. (2000). Sibling solidarity in a polygamous community in the USA: unpacking inclusive fitness. *Evolution and Human Behavior*, **21**(2), 125–39.

Lacy, R. C. and Sherman, P. W. (1983). Kin recognition by phenotype matching. *American Naturalist*, **121**, 489–512.

McLain, D. K., Setters, D., Moulton, M. P., and Pratt, A. E. (2000). Ascription of resemblance of newborns by parents and nonrelatives. *Evolution and Human Behavior*, **21**, 11–23.

Parr, L. A. and de Waal, F. B. M. (1998). Visual kin recognition in chimpanzees. *Nature*, **399**, 647–8.

Platek, S. M. (2002). An evolutionary model of the effects of human paternal resemblance on paternal investment. *Evolution and Cognition*, **9**, 1–10.

Platek, S. M., Burch, R. L., and Gallup, G. G., Jr. (2001). Sex differences in olfactory self-recognition. *Physiology & Behavior*, **73**, 635–40.

Platek, S. M., Burch, R. L., Panyavin, I., Wasserman, B., and Gallup, G. G., Jr. (2002). Children's faces: resemblance affects males but not females. *Evolution and Human Behavior*, **23**, 159–66.

Platek, S. M., Critton, S. R., Burch, R. L., *et al.* (2003) How much resemblance is enough? Determination of a just noticeable difference at which male reactions towards children's faces change from indifferent to positive. *Evolution and Human Behavior*, **23**, 81–7.

Platek, S. M., Raines, D. M., Gallup, G. G., Jr., *et al.* (2004). Reactions to children's faces: males are more affected by resemblance than females are, and so are their brains. *Evolution and Human Behavior*, **25**(6), 394–405.

Regalski, J. and Gaulin, S. (1993). Whom are Mexican infants said to resemble? Monitoring and fostering paternal confidence in the Yucatan. *Ethology and Sociobiology*, **14**, 97–113.

Reynolds, V., Falger. V. S. E., and Vine, I. (eds.) (1987). *The Sociobiology of Ethnocentrism: Evolutionary Dimensions of Xenophobia, Discrimination, Racism and Nationalism*. London: Croom Helm.

Ridley, M. (1997). *The Origins of Virtue: Human Instincts and the Evolution of Cooperation*. New York: Penguin.

Rydgren, J. (2004). The logic of xenophobia. *Rationality and Society*, **16**(2), 123–48.

Segal, N. L. and Hershberger, S. L. (1999). Cooperation and competition between twins: findings from a prisoner's dilemma game. *Evolution and Human Behavior*, **20**(1), 29–51.

Shackelford, T. K., Goetz, A. T., Buss, D. M., Euler, H. A., and Hoier, S. (2005). When we hurt the ones we love: predicting violence against women from men's mate retention tactics. *Personal Relationships*, **12**, 447–63.

Trivers, R. (1971). The evolution of reciprocal altruism. *Quarterly Review of Biology*, **46**, 35–57.

Van de Berghe, P. L. (1999). Racism, ethnocentrism, and xenophobia: in our genes or in our memes? In K. Thienpont and R. Cliquet, eds., *In-group/Out-group Behaviour in Modern Societies: An Evolutionary Perspective*. Brussels: NIDI GBGS Publications, pp. 21–33.

Van der Dennen, J. M. G. (1999). Of badges, bonds and boundaries: in-group/out-group differentiation and ethnic conflict revisited. In K. Thienpont and R. Cliquet, eds., *In-group/Out-group Behaviour in Modern Societies: an Evolutionary Perspective*. Brussels: NIDI GBGS, pp. 37–74.

Vokey, J. R., Rendall, D., Tangen, J. M., Parr, L. A., and de Waal, F. B. M. (2004). Visual kin recognition and family resemblance in chimpanzees (*Pan troglodytes*). *Journal of Comparative Psychology*, **118**(2), 194–9.

Williams, G. C. (1966). *Adaptation and Natural Selection: a Critique of Some Current Evolutionary Thought*. Princeton, NJ: Princeton University Press.

12

Children on the mind: sex differences in neural correlates of attention to a child's face as a function of facial resemblance

STEVEN M. PLATEK AND JAIME W. THOMSON
Department of Psychology, Drexel University

Introduction

Because of concealed ovulation, internal fertilization, and female infidelity, parental certainty is asymmetrical: unlike females, who are (with exception of rare maternity-room mistakes) always certain of *maternity*, males can never be certain of *paternity*. Further, during our evolutionary history, females would have benefited from 100% certainty of maternity. Current estimates of extra-pair paternity (paternity by the non-domestic father, or cuckoldry) are between 1 and 30%, with the best estimate at about 10% (Baker & Bellis, 1995; Cerda-Flores *et al.*, 1999; Neale, Neale, & Sullivan, 2002; Sasse *et al.*, 1994; Sykes & Irven, 2000). In other words, approximately 1 in 10 children are the product of female infidelity. This asymmetry in parental certainty has contributed to an asymmetry in human parental investment (Bjorklund & Shackelford, 1999 Geary, 2000). As a consequence of having to carry a child to term, females invest more in and provision more for children than do males. Additionally, if a female nurses her offspring she could be bound to a minimum of 1.5–2 years of further parental investment that is not shared by males. There are two reasons as to why parents would invest in offspring. The first is to increase their own genetic fitness (e.g. increased number of their genes in future generations) and the other is to influence their relationship with their offspring's other parent (Anderson, Kaplan, & Lancaster, 2001).

Males are not bound by their biology to provide care for offspring, and instead tend to provide care proportional to their confidence or certainty of paternity (Burch & Gallup, 2000; Daly & Wilson, 1998, 1996). The risk of cuckoldry appears to have caused the evolution of male anti-cuckoldry tactics – tactics designed to limit and control female infidelity in an attempt to reduce the likelihood of extra-pair paternity (Buss, 1988, 1994, 1999; Buss & Shackelford, 1997; Davis & Gallup, Chapter 10 in this volume, Gallup & Burch, 2004, 2005; Goetz et al., 2005; Platek, 2002; Platek et al., 2002, 2003; Shackelford et al., 2002).

We can observe a similar pattern among many other mammals. For example, paternal care is usually only manifest in those mammals with relatively high paternal certainty, whereas for most (95–97%) mammals, males provide little or no direct investment in their offspring. Among those few species that do engage in paternal provisioning, it appears that males have evolved anti-cuckoldry tactics that increase the certainty that they are the source of paternity (e.g. Lacy & Sherman, 1983). In an attempt to limit provisioning for offspring that are the consequence of female extra-pair copulations (EPCs), males of some species display what may appear to be extreme behaviors. For example, when a male Langur monkey overthrows another male and gains dominance within a troop, he will systematically kill young infants that were fathered by the previous alpha male. By resorting to *infanticide* when the paternity of an offspring is ostensibly foreign (e.g. Hrdy, 1974), his behavior serves two adaptive functions: (1) it eliminates the possibility that he will invest valuable resources in unrelated offspring and (2) it induces menstrual cycling (i.e. sexual receptivity) in those females whose offspring he killed. This allows the new dominant male to use the females to his own reproductive advantage. Additionally, male baboons appear to invest resources in offspring proportional to the degree to which they monopolized the female prior to insemination (Buchan et al., 2003).

There is growing evidence that human males are similarly affected by these evolutionary pressures to invest in offspring as a function of paternal certainty. As a way of elucidating the importance of paternity for males, Daly and Wilson (1982) and Regalski and Gaulin (1993) observed families in maternity wards. They measured the number of times people remarked whether the infant looked more like the mother or father. Both studies found that people were more likely to ascribe resemblance to putative fathers than they were to ascribe resemblance to mothers. Both studies also documented that the mother and her family were more likely to attribute resemblance to the putative father than were other people. They interpreted these behaviors as attempts by the female and her family to convince the putative father that he sired the child.

It is also well known that men differentially invest resources in children to whom they are genetically related. For example, it is not uncommon for unrelated or stepchildren, to be treated significantly worse than biological children (e.g. Anderson *et al.*, 1999). Previous research has shown that children raised in stepfamilies are more disadvantaged than children raised in single-mother homes when it comes to things like educational achievement, steady employment in early adulthood, stable marriages, and mental health (see Case *et al.*, 2001). Burch and Gallup (2000) have shown that males spend less time with, invest fewer resources in, and are more likely to abuse ostensibly unrelated children than children they assume to be their genetic offspring. They also found that the less a male thinks a child looks like him regardless of actual genetic relationship to the child, the worse he treats the child and the worse the relationship with that child is. Daly and Wilson (1988a; and see Daly, Wilson, & Weghorst, 1982) estimate the incidence of abuse that results in infanticide among stepchildren to be as great as 100 times that of genetically related children. In Daly and Wilson's (1988b) landmark book, *Homicide*, they interpret spousal homicide (uxorocide) as a byproduct of cuckoldry fear and sexual jealousy among men. These data suggest a strong link between paternity uncertainty and family violence.

As a result of paternal uncertainty, human males have evolved an arsenal of anti-cuckoldry tactics to limit and control the incidence of female infidelity, thereby increasing the likelihood that the children they provision are genetically related to them. Emerging data suggest that males have evolved at least three types of tactic that help to reduce the likelihood of being cuckolded: (1) monitoring/mate guarding his partner during the fertile period (Daly & Wilson, 1998), (2) intra-vaginal inter-male competition, such as sperm competition (Birkhead, 1995, 1996; Shackelford *et al.*, 2002) and semen displacement (Gallup *et al.*, in press-d), and (3) assessing paternity post-parturition (see also Chapter 7 in this volume). Daly and Wilson (1998) and others (Platek *et al.*, 2002, 2003; Platek, 2002) have hypothesized that males infer paternity by assessing physical resemblance.

Assessing paternal resemblance

An indirect way that resemblance might affect male behavior toward children is by a "social mirror" (Burch & Gallup, 2000); that is, other people can ascribe paternal resemblance to the child as a means of swaying a male to act more positively towards the child. Daly and Wilson (1982) and Regalski and Gaulin (1993) have shown that mothers and family members actively ascribe paternal resemblance to children and that when males express doubt they are quick to reassure them of resemblance. Males may be predisposed to take into account social-mirror information because of the importance of paternity (see

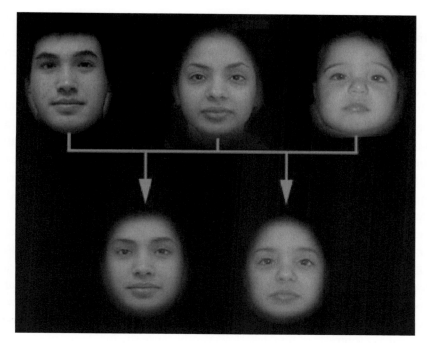

Figure 12.1. Example of morphing paradigm. An unknown adult face (upper left face) is morphed with subject's face (upper center face) to create an adult composite adult morph (bottom left face) and an unknown child's face (top right face) is morphed with a subject's face to create a composite child morph (bottom right face).

Hauber & Sherman, 2001; Neff & Sherman, 2002 for reliable/unreliable cues of parentage; i.e. paternity).

Burch and Gallup (2000) found that the more a sample of convicted spouse abusers felt that their children looked like them, the better the children were treated. The recalled childhoods of the abusive males themselves also were affected by how much they thought they resembled their fathers. Perceptions of paternal resemblance were negatively correlated with the recalled incidence of physical and sexual abuse they experienced as children, as well as feelings of closeness to their father. How often others had told them that the child physically resembled them also correlated with a male's ratings of his relationship to the child.

A male, however, also may assess the degree to which a child actually resembles him. Platek *et al.* (2002) morphed the faces of participants (see Figure 12.1) with the faces of toddlers and measured reactions to hypothetical investment questions (e.g. Which one of these children would you spend the most time with?, Which one would you adopt?). Males were more likely than females, and more likely than chance, to select for investment a child whose face their own face had been morphed with. Thus it seems that actual resemblance also plays a role in a male's reactions toward children's faces and this might be modulated by a cortical mechanism or module (e.g. self-referent phenotype matching; see Neff & Sherman, 2002) dedicated to controlling the affective nature of males' reactions toward children.

In a test of how actual resemblance and social-mirror-mediated resemblance interact to effect reactions toward children's faces, Platek (2002; see also Chapter 11 in this volume) provided participants with social-mirror information about children's faces, some of which were morphed to resemble the participant, and found that social-mirror information affected male and female reactions similarly. However, unlike females, males were still more affected by actual resemblance; males selected a face primarily based on whether the face resembled them. This study replicates previous findings that males use actual resemblance in their reactions toward hypothetical children (Platek *et al.*, 2002, 2003) and supports the hypothesis that social perceptions of resemblance also affect father–child relationships (Burch & Gallup, 2000). However, the relative importance of actual versus socially perceived resemblance is not yet known.

Convincing a male of paternity and securing his investment would be in the evolutionary interests of females. However, because the incidence of cuckoldry is appreciable (1–30% as noted above), it is hardly in the interests of males to be convinced easily. If ascriptions of resemblance were fully persuasive throughout evolutionary history, males might have been deceived into investing in children that were sired by rival males. Those who remained wary and used their own perceptions of resemblance and invested accordingly likely would have been more successful reproductively.

McLain *et al.* (2000) investigated this idea by comparing new mothers' ascriptions of paternal resemblance to the ability of independent raters to match photos of children to their putative fathers. Maternal ascriptions of resemblance could not be verified by the objective, unrelated raters. In no case were the infants' pictures matched to the fathers' photos more often than chance; the mothers' opinions, although highly reliable, held no validity. These findings might, in turn, explain the reluctance of males to quickly agree with their partner's assertions of paternity (Daly & Wilson, 1982; Regalski & Gaulin, 1993). In some cases, males agree to such ascriptions after several maternal attempts to persuade them. Interesting as these data are, they are still flawed in that actual paternity was never determined, which could have masked or obscured independent raters' ability to match children to the males (see also Brédart & French, 1999; Christenfeld & Hill, 1995; Nesse, Silverman, & Bortz, 1990).

In addition, a male may have adopted a strategy of comparing their offspring to their kin in order to assess resemblance and also choose to believe information provided only by those who also shared genes in common with him. Platek *et al.* (2003) and DeBruine (2002, 2004) have provided indirect evidence that parentage and trust may also be mediated by facial resemblance. Platek *et al.* (2003; but see DeBruine, 2004) found that males react favorably toward children's faces that shared 25% of their characteristics, which is approximately the proportion of genes shared in common with kin one step removed: grandchildren, nieces and nephews, aunts and uncles, and half-siblings. DeBruine (2002) found that participants tended to trust faces that resembled them more than those that did not and has since shown a self-resemblance attractiveness bias (L. M. De Bruine, personal communication).

In all of our studies to date (Platek *et al.*, 2002, 2003; Platek, 2002; Platek, unpublished data), none of the participants was aware of the effect resemblance had on their choices. In a re-analysis of existing data from our previous morphing studies (Platek *et al.*, 2002, 2003; Platek, 2002) and in conjunction with data not yet published, no participant reported using resemblance to choose a child's face. When queried about their choices, none of the subjects identified resemblance as a factor in how they chose which child to select, and neither did they realize that their faces had been morphed with the child. In fact, during debriefing when subjects were told about the hypothesis, and shown the morphing procedure, most subjects responded with surprise and asked to see the faces again in an attempt to identify consciously which face it was that their own face had been morphed with. Even under these conditions, participants still had difficulty selecting the face that their own face had been morphed with. It was not until their original picture was available for comparison to the child morphs that participants could tell which face their own face been morphed with. This suggests that males posses a mechanism that processes information about resemblance at largely unconscious levels. Therefore, it has been hypothesized (Platek *et al.*, 2004a; Platek, Keenan, and Mohamed, 2005) that male brains may support neural architecture for a resemblance-detection module that is situation (child-care)-specific.

Neurobiological correlates

If the sex difference in reaction to children's faces is produced by evolutionary pressures and is happening at levels below conscious awareness, one might expect the coevolution of (sex-) specialized modules for processing such information. Questions about the neurobiological correlates of adaptations have only recently been posed and investigated (Duchaine, 2002; Platek *et al.*, 2004a, 2005).

Here we briefly outline the methods of functional magnetic resonance imaging (fMRI) and how our research group has used this method to investigate sex differences in perceptions of facial resemblance.

METHODS: fMRI
Blood Oxygenation Level Dependent (BOLD) signal overview

fMRI (Ogawa et al., 1992) of the brain using the state-of-art echo planar imaging (EPI) provides images of brain activity during a task (e.g. motor, visual, cognitive, etc.). The BOLD contrast mechanism is the basis of MRI that was used in this study. During neural firing it is hypothesized that an increase in partial pressure of oxygen in blood causes an increase in hemoglobin oxygen saturation, decreasing the concentration of paramagnetic deoxyhemoglobin. This decrease in the concentration of deoxyhemoglobin causes a reduction in the tissue–blood susceptibility differential which, in turn, decreases spin dephasing and results in an increased $T2^*$ signal; that is, allowing the MRI scanner to detect subtle differences in signal intensity. Using correlated changes in blood oxygenation and cognitive tasks one can make inferences about underlying neuronal activity and the specific brain substrates implicated in specific perceptual, motor, and cognitive processes. The BOLD response is the most common means for mapping human brain–behavior relationships in vivo.

Analysis of fMRI data

The data collected from fMRI studies are extremely complex and need pre- and post-processing in order to accurately interpret the changes in activation associated with a specific cognitive task(s). The methods for doing these procedures are outlined in a number of excellent texts (e.g. Bandettini, 2002; Frackowiak et al., 2003; see also http://www.fil.ion.ucl.ac.uk/spm/doc/books/hbf2/) and papers (e.g. Ashburner & Friston, 1999; Friston et al., 1995). The post-acquisitional spatial and statistical selections of the analysis are typically performed using mathematical software packages such as SPM (Statistical Parametric Mapping; Wellcome Department of Cognitive Neurology, University College London, London, UK), run under the Matlab® (The Mathworks, Natick, MA, USA) environment or FSL (FMRIB, Oxford, UK), which is available for any operating system.

Activation mapping

To determine the spatial extent of the fMRI activation and to subsequently determine the BOLD signal, the general linear model (GLM) is applied to the data. A random effects analysis is employed to determine group-level statistical parametric maps (SPMs) that are considered generalizable to a population.

Table 12.1. *Brain-activation regions (Talairach coordinates) when comparing activation between self-child morphs and non-self-child morphs in (a) males and (b) females.*

(a)

| Right hemisphere | Coordinates | | | BA | P value |
	x	y	z		
Middle frontal gyrus	−28	44	18	10	<0.05
Superior frontal gyrus	−30	54	14	10	<0.05

(b)

| Right hemisphere | Coordinates | | | BA | P value |
	x	y	z		
Inferior frontal gyrus	60	14	16	44/45	<0.01
Middle frontal gyrus	46	10	30	9	<0.01
Left medial frontal lobe					
Medial frontal gyrus	−12	50	10	10	<0.01
Medial Superior frontal gyrus	−18	60	8	–	<0.01

BA, Broadman's area.

FINDINGS FOR SEX DIFFERENCES IN BRAIN ACTIVATION TO FACIAL RESEMBLANCE

Two recent studies by our research group have provided initial evidence in favor of a sex-specific modular response to facial resemblance as a function of age. In the first of these studies, Platek *et al.* (2004a) asked nine participants to look at facial images of children, some of which had been morphed with the subject's face. Two interesting findings emerged from this initial study: first, we found that males showed activation in the anterior left prefrontal lobe and anterior cingulate gyrus (see Figure 12.1 and Table 12.1a) when viewing children who were morphed to resemble them, but females did not; and, second, we found that females activated a set of substrates in the right and medial prefrontal cortices when viewing all children's faces, irrespective of whether the face was morphed with theirs (see Figures 12.2 and 12.3, and Table 12.1b).

In order to extend and improve upon this study, we (Platek *et al.*, 2005) designed a more rigorous study including a control condition for adult faces. Not only did we control for exposure to adult faces, but we also used a more

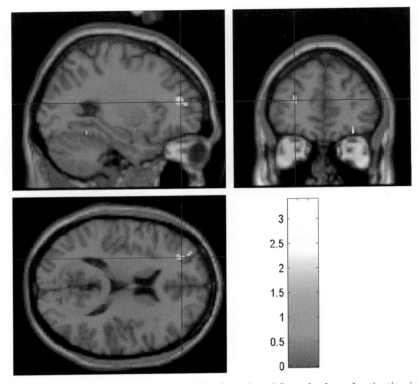

Figure 12.2 Left superior and middle frontal, and frontal subgyral activation in males when comparing activation between self-child morph and non-self child morph conditions. The bar shows the value of the *t*-statistic; $P < 0.005$.

sensitive experimental paradigm (known as event-related fMRI). The event-related design is superior in measuring hemodynamic response or changes over time and allows more experimental flexibility to create ecologically valid experiences for the subject. Participants were then asked to actively respond to the faces as they were presented in the scanner. This study generated results consistent with the previous research. We found that male brains were much more active when responding to children's faces that resembled them when compared to responding to either other children's faces or adult faces who did and did not resemble them, and also showed more activation than females to children's faces that resembled them. When comparing activation between responding to children's faces that resembled the participant and children's faces that did not resemble the participant, we found significant activation in the anterior cingulate and anterior left prefrontal lobe in males and caudate nucleus in females (see Figure 12.4 and Table 12.2).

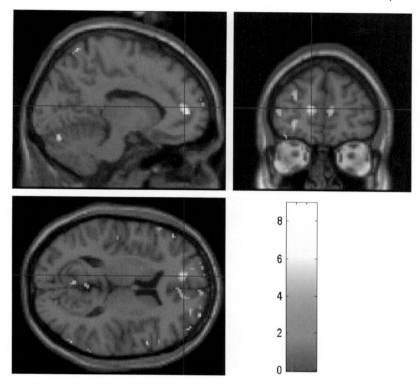

Figure 12.3 Left medial and medial superior frontal gyrus activation in females when comparing activation between self-child morph and non-self child morph conditions. The bar shows the value of the t-statistic; $P < 0.005$.

General discussion and conclusions

The combined data from the two experiments described above suggest that the behavioral difference that males and females show in reactions toward children's faces (Burch & Gallup, 2000; Daly & Wilson, 1998; Platek, 2002; Platek *et al.*, 2002, 2003, 2004a) is associated with differences in neural processing (Platek *et al.*, 2004a, 2005). Unlike females, males showed significant neural activation in the left frontal cortex and anterior cingulate region, which has been hypothesized to be involved in inhibition of negative responses (Collette *et al.*, 2001; Davidson, 1997; Harmon-Jones & Sigelman, 2001). Applying this hypothesis to these data, it would appear that males may possess a generalized skepticism about children that is inhibited when (1) the child resembles him and (2) he is faced with the adaptive problem of provisioning for offspring. Thus the left frontal activation associated with viewing self-child morphs may be a situation-specific evolutionarily adaptive response in males, which supports the

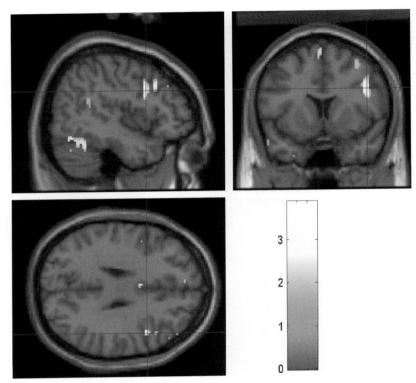

Figure 12.4 Right middle and inferior frontal lobe activation in females when comparing activation between self-child morph and non-self child morph conditions. The bar shows the value of the *t*-statistic; $P < 0.005$.

hypothesis put forth by Daly and Wilson (1998) and Platek and colleagues (Platek *et al.*, 2002, 2003; Platek, 2002) that males use self-resemblance to assess paternity.

Additionally, we found (Platek *et al.*, 2005) that males showed more overall activation to faces that resembled them when compared to females. This finding suggests that males might possess an inherent advantage in detection of resemblance among both children and adult faces; that is, the male brain may have been selected to be more sensitive to facial resemblance. In the realm of dissortative mating, this may aid males in selection of genetically compatible mates, thereby limiting the incidence of inbreeding depression due to homogeneity by descent. However, this hypothesis is inconsistent with recent evidence that shows females are slightly better at pairing couples based on facial resemblance (Alvarez & Jaffe, 2004). However, women's advantage in pairing couples based on resemblance might be an expression of their desire to enter into relationships with males who resemble them as a means with which to

Table 12.2. *Brain-activation regions (Talairach coordinates) when comparing activation between self-child morphs and non-self-child morphs in (a) males and (b) females using an event-related fMRI experimental design.*

Region	hemisphere	Coordinates			Z score
		x	*y*	*z*	
(a)					
Male > female					
Superior temporal gyrus	Right	28	14	−24	4.54
Cingulate gyrus	Left	−18	26	23	4.27
Inferior parietal lobe	Right	46	−45	41	4.03
Precuneus	Right	6	−62	45	4.01
Anterior cingulate gyrus	Left	0	34	17	3.88
Anterior cingulate gyrus	Left	−2	−18	32	3.49
(b)					
Female > male					
Insula	Left	−40	−5	13	3.52
Cingulate gyrus	Left	0	25	37	3.46

BA, Broadman's area.

capitalize on paternal resemblance strategies. In other words, by pair-bonding with a mate that resembles her, a female might be better equipped to confuse paternity through extra-pair mating. If children share resemblance with mothers, and they likely do, then by pair-bonding with a male who resembles oneself a female is in a unique position to use shared resemblance in her mate to confuse paternity. That is, a female who pair-bonds with a male who resembles her might be less likely to incur the negative consequences of detected extra-pair mating because any offspring will share genes in common with her, and therefore share some resemblance with her, and possibly share resemblance with her mate, because she had selected her mate based upon his resemblance to her. Thus, we would predict that women would prefer (find more sexually attractive) the faces of males that resemble them as a consequence of an adaptive strategy for confusing in-pair paternity, not as a consequence of increased vanity. On the other hand, males would be predicted to prefer (find more sexually attractive) female faces that did not resemble them as a means to increase possible detection of paternity.

The female data in our study are harder to explain. However, in light of anecdotal behavioral evidence, it might be the case that the right lateralized frontal and medial frontal activity may be part of a mentalizing (e.g. theory of mind or mental state attribution) module. In each of our behavioral studies (e.g. Platek *et al.*, 2002), participants were queried as to how they made their choices. Whereas males usually reported "going with a gut feeling" or using no strategy at all, females often would attribute specific personality characteristics to the children's faces. Females reported that they tried to find the "nicest" child to give money to, spend time with, or adopt, and the "meanest" or "brattiest" child to punish, not spend time with, and not spend money on.

Females may be making decisions to invest in or discipline children based on inferences about psychological characteristics of the child. This may be supported by recent neuropsychological and neuroimaging data on mentalizing (inferring the psychological characteristics of another person). For example, Fletcher *et al.* (1995) found medial prefrontal cortex (MPFC) activation when comparing activation associated with reading mentalizing stories and reading physical stories. Gallagher *et al.* (2000) and Vogely *et al.* (2001) found similar MPFC, as well as right-hemisphere, activation associated with mental state processing. In a test similar to that of Povinelli, Rulf, & Bierschwale (1994) for understanding intention in chimpanzees, Berthoz *et al.* (2002) reported MPFC activation associated when reading social transgressions that were both deliberate and accidental. These data extend those of Castelli *et al.* (2000) and Klin *et al.* (2003) which demonstrated activation in the temporal pole and MPFC when observing motion of inanimate objects that could be interpreted as having intention or desire. Stuss and colleagues (Ishii *et al.*, 2002; Stuss, Gallup, & Alexander, 2001) showed that patients with damage to the right frontal lobes, but not other parts of the cortex, were deficient at understanding visual perspective taking, deception, and emotional mental states. Using fMRI, Platek *et al.* (2004b) showed activation in the medial and right frontal lobes when participants were asked to think about the mental states of others when seeing only the person's eyes (Eye in the Mind-Revised Test; Baron-Cohen *et al.*, 2001). Therefore, our finding that females showed activation in right and medial prefrontal substrates and previous anecdotal reports that females tended to use more personality-based characteristics to make behavioral decisions supports the notion that females may utilize a mentalizing approach when thinking about how to invest resources on children. Why the sex difference in approach to children exists is still unclear. However, it might be that males are reacting based on paternity-uncertainty mechanisms, whereas females are evaluating children and their investment in children much the way they might for adults; that is, based upon whether they like/dislike the person.

There are obvious limitations to this study. For example, the sample size is small; even though large sample sizes are not typically needed to achieve adequate statistical power in functional imaging studies, five subjects is a smaller than average (which is usually 8–12, or more). Additionally, a box-car design was used in our first study, which is not as sensitive to subtle changes in neural activations as event-related fMRI and one runs the risk of habituation effects. However, in order to account for the possibility of low activity levels and possibility of habituation, we used six blocks, which has been shown to be a reliable number of epochs to produce maximum activation associated with stimulus exposure while limiting the likelihood of habituation effects (F. B. Mohamed, personal communication).

In order for successful mating to occur, an individual must be able to resolve the adaptive problem of retention. While this is relevant to both sexes, there are increased costs for males who are not successful in the mating process. Mate-guarding tactics have evolved in species that participate in social monogamy and enabled males, among other things, to help ensure parental certainty so that they do not waste valuable resources on offspring that do not share any genes in common. Female infidelity has led to the adaptive process of sperm competition and it became necessary for males to (1) prevent a partner's infidelity, (2) correct a partner's infidelity, and (3) anticipate a partner's infidelity (Shackelford, 2003). Solving these adaptive problems may have led males to develop such mechanisms that would allow them to be more successful in recognizing kin under situations of child rearing, which would translate into increased reproductive fitness.

Acknowledgments

The authors thank Feroze Mohamed and Scott Faro for their assistance with the functional magnetic resonance imaging. We also thank Gordon Gallup, Rebecca Burch, Julian Keenan, Todd Shackelford, Thomas Myers, Samuel Critton, Ivan Panyavin, Brett Wasserman, Gordon Bear, David Smith, and Robert Haskell for helpful discussion concerning this line of investigation.

References

Alvarez, L. and Jaffe, K. (2004). Narcissism guides mate selection: humans mate assortatively, as revealed by facial resemblance, following an algorithm of "self seeking like". *Evolutionary Psychology*, **2**, 177–94.

Anderson, K., Kaplan, H., Lam, D., and Lancaster, J. (1999). Paternal care of genetic fathers and stepfathers II: reports by Xhosa high school students. *Evolution and Human Behavior*, **20**, 433–51.

Anderson, K., Kaplan, H., and Lancaster, J. (2001). Men's financial expenditures on genetic children and stepchildren from current and former relationships. *PSC Research* (Population Studies Center), Report No. 01–484.

Ashburner, J. and Friston, K. J. (1999). Nonlinear spatial normalization using basis functions. *Human Brain Mapping*, **7**(4): 254–66.

Baker, R. R. and Bellis, M. A. (1995). *Human Sperm Competition: Copulation, Masturbation, and Infidelity*. London: Chapman and Hall.

Bandettini, P. A. (2002). Functional MRI. In F. Bolles and J. Grofman, eds., *Handbook of Neuropsychology*. Amsterdam: Elsevier.

Baron-Cohen, S., Wheelright, S., Hill, J., Raste, Y., and Plumb, I. (2001). The "Reading the Mind in the Eyes" test revised version: a study with normal adults, and adults with Asperger syndrome or high-functioning autism. *Journal of Child Psychology and Psychiatry*, **42**, 241–51.

Berthoz, S., Armony, J. L., Blair, R. J., and Dolan, R. J. (2002). An fMRI study of intentional and unintentional (embarrassing) violations of social norms. *Brain*, **125**, 1696–708.

Birkhead, T. R. (1995). Sperm competition: evolutionary causes and consequences. *Reproduction, Fertility, and Development*, **7**, 755–75.

Birkhead, T. R. (1996). Sperm competition: evolution and mechanisms. *Current Topics in Developmental Biology*, **33**, 103–58.

Bjorklund, D. F. and Shackelford, T. K. (1999). Differences in parental investment contribute to important differences between men and women. *Current Directions in Psychological Science*, **8**(3), 86–9.

Brédart, S. and French, R. (1999). Do babies resemble their fathers more than their mothers? A failure to replicate Christenfeld and Hill. *Evolution and Human Behavior*, **20**, 129–35.

Buchan, J. C., Alberts, S. C., Silk, J. B., and Altmann, J. (2003). True paternal care in a multi-male primate society. *Nature*, **425**, 179–81.

Burch, R. L. and Gallup, G. G., Jr. (2000). Perceptions of paternal resemblance predict family violence. *Evolution and Human Behavior*, **21**, 429–35.

Buss, D. M. (1988). From vigilance to violence: tactics of mate retention in American undergraduates. *Ethology and Sociobiology*, **9**, 291–317.

Buss, D. M. (1994). *The Evolution of Desire*. New York: Basic Books.

Buss, D. M. (1999). *Evolutionary Psychology*. Needham Heights, MA: Allyn & Bacon.

Buss, D. M. and Shackelford, T. K. (1997). From vigilance to violence: mate retention tactics in married couples. *Journal of Personality and Social Psychology*, **72**, 346–61.

Case, A., Lin, I. F., and McLanahan, S. (2001). Educational attainment of siblings in stepfamilies. *Evolution and Human Behavior*, **22**, 269–89.

Castelli, F., Happe, F., Frith, U., and Frith, C. D. (2000). Movement and mind: a functional imaging study of perception and interpretation of complex intentional movement patterns. *NeuroImage*, **12**, 314–25.

Cerda-Flores, R. M., Barton, S. A., Marty-Gonzales, L. F., Rivas, F., and Chakrborty, R. (1999). Estimation of nonpaternity in the Mexican population of Nuevo Leon: a validation study with blood group markers. *American Journal of Physical Anthropology*, **109**, 281–93.

Christenfeld, N. and Hill, E. (1995). Whose baby are you? *Nature*, **378**, 669.

Collette, F., Van der Linden, M., Delfiore, G., *et al.* (2001). The functional anatomy of inhibition processes investigated with the Hayling Task. *NeuroImage*, **14**, 258–67.

Daly, M. and Wilson, M. (1982). Whom are newborn babies said to resemble? *Ethology and Sociobiology*, **3**, 69–78.

Daly, M. and Wilson, M. (1988a). Evolutionary social psychology and family homicide. *Science*, **242**, 519–24.

Daly, M. and Wilson, M. (1988b). *Homicide*. Hawthorne NY: Aldine de Gruyter.

Daly, M. and Wilson, M. (1996). Violence against stepchildren. *Current Directions in Psychological Science*, **5**, 77–81.

Daly, M. and Wilson, M. (1998). *The Truth About Cinderella: a Darwinian View of Parental Love*. New Haven, CT: Yale University Press.

Daly, M., Wilson, M., and Weghorst, S. J. (1982). Male sexual jealousy. *Ethology and Sociobiology*, **3**(1), 11–27.

Davidson, R. J. (1997). Emotion and affective style: physiological substrates. *Electroencephalography and Clinical Neurophysiology*. American Encephalographic Society, Abstract 102.

DeBruine, L. M. (2002). Facial resemblance enhances trust. *Proceedings of the Royal Society of London B*, **269**, 1307–12.

DeBruine, L. M. (2004). Resemblance to self increases the appeal of child faces to both men and women. *Evolution and Human Behavior*, **25**, 142–54.

Duchaine, B. (2002). *Computational and developmental specificity in face recognition*. Doctoral dissertation, University of California, Santa Barbara. *Dissertation Abstracts International*, **62**, 3821B.

Fletcher, F. C., Happe, F., Frith, U., *et al.* (1995). Other minds in the brain: a functional imaging study of "theory of mind" in story comprehension. *Cognition*, **57**, 109–28.

Frackowiak, R. S. J., Friston, K. J., Frith, C., *et al.* (2003). *Human Brain Function*, 2nd edn. New York: Academic Press.

Friston, K. J., Holmes, A. P., Worsley, K. J., *et al.* (1995). Statistical parametric maps in functional imaging: a general linear approach. *Human Brain Mapping*, **2**, 189–210.

Gallagher, H. L., Happe, F., Brunswick, N., *et al.* (2000). Reading the mind in cartoons and stories: an fMRI study of 'theory of mind' in verbal and nonverbal tasks. *Neuropsychologia*, **38**, 11–21.

Gallup, Jr., G. G. and Burch, R. (2004) Semen displacement as a sperm competition strategy in humans. *Evolutionary Psychology*, **2**, 12–23.

Gallup, G. G., Jr. and Burch, R. L. (2005). The semen displacement hypothesis. In T. Shackelford and N. Pound eds., *Sperm Competition in Humans: Classic and Contemporary Readings*. Berlin: Springer.

Geary, D. C. (2000). Evolution and proximate expression of human paternal investment. *Psychological Bulletin*, **126**, 55–77.

Goetz, A. T., Shackelford, T. K., Weekes-Shackelford, V. A., *et al.* (2005). Mate retention, semen displacement, and human sperm competition: a preliminary investigation of tactics to prevent and correct female infidelity. *Personality and Individual Difference*, **38**, 749–63.

Harmon-Jones, E. and Sigelman, J. (2001). State anger and prefrontal brain activity: evidence that insult-related relative left-prefrontal activation is associated with experienced anger and aggression. *Interpersonal Relations and Group Processes*, **5**, 797–803.

Hauber, M. E. and Sherman, P. W. (2001). Self-referent phenotype matching: theoretical considerations and empirical evidence. *Trends on Cognitive Sciences*, **10**, 609–16.

Hrdy, S. (1974). Male-male competition and infanticide among the langurs (*Presbytis entellus*) of Abu, Rajasthan. *Folia Primatologica (Basel)* **22**(1), 19–58.

Ishii, R., Stuss, D. T., Gorjmerac, C., *et al.* (2002). Neural correlates of "theory of mind" in emotional vignettes comprehension studied with spatially filtered magnetoencephalography. In Nowak, H., Haueisen, J., Giessler, F., and Hounker, R., eds., *Biomag 2002 Proceedings of the 13th International Conference on Biomagnetism* Berlin: VDE Verlag GMBH, 291–3.

Klin, A., Jones, W., Schultz, R., and Volkmar, F. (2003). The enactive mind, or from actions to cognition: lessons from autism. *Philosophical Transactions of the Royal Society of London B*, **358**, 345–60.

Lacy, R. C. and Sherman, P. W. (1983). Kin recognition by phenotype matching. *American Naturalist*, **121**, 489–512.

McLain, D. K., Setters, D., Moulton, M. P., and Pratt, A. E. (2000). Ascription of resemblance of newborns by parents and nonrelatives. *Evolution and Human Behavior*, **21**, 11–23.

Neale, M. C., Neale, B. M., and Sullivan, P. F. (2002). Nonpaternity in linkage studies of extremely discordant sib pairs. *American Journal of Human Genetics*, **70**, 526–9.

Neff, B. D. and Sherman, P. W. (2002). Decision making and recognition mechanisms. *Proceedings of the Royal Society of London B*, **269**, 1435–41.

Nesse, R., Silverman, A., and Bortz, A. (1990). Sex differences in ability to recognize family resemblance. *Ethology and Sociobiology*, **11**, 11–21.

Ogawa, S., Tank, D. W., Menon R., *et al.* (1992). Intrinsic signal changes accompanying sensory stimulation: functional brain mapping with magnetic resonance imaging. *Proceedings of the National Academy of Science USA*, **89**, 5951–5.

Platek, S. M. (2002). Unconscious reactions to children's faces: the effect of resemblance. *Evolution and Cognition*, **8**, 207–14.

Platek, S. M., Burch, R. L., Panyavin, I. S., Wasserman, B. H., and Gallup, G. G., Jr. (2002). Reactions towards children's faces: resemblance matters more for males than females. *Evolution and Human Behavior*, **23**, 159–66.

Platek, S. M., Critton, S. R., Burch, R. L., *et al.* (2003). How much paternal resemblance is enough? Sex differences in the reaction to resemblance but not in ability to detect resemblance. *Evolution and Human Behavior*, **24**, 81–7.

Platek, S. M., Raines, D. M., Gallup, Jr., G. G., *et al.* (2004a). Reactions to children's faces: males are still more affected by resemblance than females are, and so are their brains. *Evolution and Human Behavior*, **25**, 394–405.

Platek, S. M., Keenan, J. P., Gallup, G. G., Jr., and Mohamed, F. B. (2004b). Where am I? Neural correlates of self and other. *Cognitive Brain Research*, **19**, 114–22.

Platek, S. M., Keenan, J. P., and Mohamed, F. B., (2005). Neural correlates of facial resemblance. *NeuroImage*, **25**, 1336–44.

Povinelli, D. J., Rulf, A. B., and Bierschwale, D. T. (1994). Absence of knowledge attribution and self-recognition in young chimpanzees (*Pan troglodytes*). *Journal of Comparative Psychology*, **108**, 74–80.

Regalski, J. and Gaulin, S. (1993). Whom are Mexican infants said to resemble? Monitoring and fostering paternal confidence in the Yucatan. *Ethology and Sociobiology*, **14**, 97–113.

Sasse, G., Muller, H., Chakraborty, R., and Ott, J. (1994). Estimating the frequency of nonpaternity in Switzerland. *Human Heredity*, **44**, 337–43.

Shackelford, T. (2003). Preventing, correcting and anticipating female infidelity: three adaptive problems of sperm competition. *Evolution and Cognition*, **9**, 90–6.

Shackelford, T. K., LeBlanc, G. J., Weekes-Schakelford, V. A., *et al.* (2002). Psychological adaptations to human sperm competition. *Evolution and Human Behavior*, **23**, 123–38.

Stuss, D. T., Gallup, G. G., Jr., and Alexander, M. P. (2001). The frontal lobes are necessary for theory of mind. *Brain*, **124**, 279–86.

Sykes, B. and Irven, C. (2000). Surnames and the Y chromosome. *American Journal of Human Genetics*, **66**, 1417–9.

Vogely, K., Bussfeld, P., Newen, A., *et al.* (2001). Mind reading: neural mechanisms of theory of mind and self-perspective. *NeuroImage*, **14**, 170–81.

Index